Frank Straube · Shihua Ma · Michael Bohn

Internationalisation of Logistics Systems

Frank Straube · Shihua Ma · Michael Bohn

Internationalisation of Logistics Systems

How Chinese and German companies
enter foreign markets

Springer

Frank Straube
TU Berlin
Strasse des 17. Juni 135
10623 Berlin
Germany
straube@logistik.tu-berlin.de

Michael Bohn
TU Berlin
Strasse des 17. Juni 135
10623 Berlin
Germany
bohn@logistik.tu-berlin.de

Shihua Ma
Huazhong University of Science & Technology
Wuhan
430074 Hubei
P.R. China
shihuama@public.wh.hb.cn

ISBN 978-3-540-76982-8   e-ISBN 978-3-540-76984-2

DOI 10.1007/978-3-540-76984-2

Library of Congress Control Number: 2007939801

Cover Design: WMXDesign GmbH, Heidelberg

Printed on acid-free paper

9 8 7 6 5 4 3 2 1

springer.com

Competence Center for International Logistics Networks, Berlin, Germany

The Competence Center for International Logistics Networks was established as part of the Logistics Department at the Berlin University of Technology (TUB) in 2005, and is financed by the Swiss Kuehne-Foundation. Its purpose is to carry out research in the field of global logistics in interaction with international partners from academia and business practice. The Competence Center is headed jointly by Prof. Dr.-Ing. Frank Straube and Prof. Dr.-Ing. Helmut Baumgarten. Prof. Dr.-Ing. Frank Straube is Director of the TUB Logistics Department and Managing Director of the TUB Institute of Technology and Management. Furthermore, he is Vice President of the German Logistics Association (BVL) and member of the Board of the European Logistics Association (ELA). Being an active member of the European logistics community, he published extensively on the subject of logistics and is a renowned presenter on national and international conferences. Dipl.-Ing. oec. Michael Bohn studied Industrial Engineering and Management in Hamburg and Singapore, and gained experience in the manufacturing industry and consulting business, before he entered the TUB Logistics Department in 2005. He works as Research Associate at the Competence Center for International Logistics Networks, specialising on research in the field of global logistics and he is managing international research projects.
For further details please visit: www.internationalelogistik.de

Institute of Supply Chain & Logistics Management, Wuhan, China

The Institute of Supply Chain & Logistics Management (ISC&LM) in Wuhan was founded at the Huazhong University of Science and Technology (HUST) in 1979. It is one of the earliest institutions active in research in logistics and supply chain management in China. In recent years, the ISC&LM has undertaken extensive international cooperation with overseas universities. The head of the ISC&LM is Professor Ma Shihua, who is also the Vice Dean of the School of Management at the HUST. He has been engaged in research and industrial practice in supply chain and logistics management for many years, as well as in production and operations management. His institute offers several courses including supply chain management, operations management and logistics management for EMBA, MBA, graduate and undergraduate students. He has consulting experience with numerous companies in China.
For further details please visit: www.logistics-chain.com

Authors

Prof. Dr.-Ing.
Frank Straube
Berlin University of Technology

Prof.
Shihua Ma
Huazhong University (HUST)

Dipl.-Ing. oec.
Michael Bohn
Berlin University of Technology

Research Associates

Dipl.-Inform.Wirt
Daniel Rief
Berlin University of Technology

Dipl.-Ing.
Jörg Pohl
Berlin University of Technology

Dr.
Kunpeng Li
Huazhong University (HUST)

Student Assistance

- Roman Figiel
- Sebastian Flucht
- Pravin Narkhedkar
- Jia Qu
- Christopher Schwarz
- Kuo-Hsiang Yong
- Xin Zhang
- Kun Huang
- Huijun Li
- Jing Xin
- Chen Wei
- Xu Guan
- Binzong Peng
- Zheng Lei
- Zheng Huang

Layout

- Axel Haas

Our Dearest Reader,

almost every company today is in one way or another part of a network of global supply chains all relying critically on efficient and reliable logistics systems. For enterprises of any size, the phenomenon of globalisation creates a multitude of opportunities to enter new regional markets, and equally requires a multitude of enablers. These form the core of the research on the **Internationalisation of Logistics Systems**. And it is the wealth of opportunities afforded by the interaction between Germany and China—the export world champion and the world's fastest developing economy—that the focus of this survey is on these two global players.

The survey determines the logistics strategies that Chinese and German companies apply when expanding globally. Seven of the most important target regions for internationalisation activities are covered by this survey. The specific challenges and opportunities in the different markets are determined from a logistics perspective, thereby comparing how Chinese and German companies manage to become successful.

The research was conducted by the Competence Center for International Logistics Networks, which is part of the Logistics Department at the Berlin University of Technology (TUB) in close cooperation with the Supply Chain & Logistics Institute at the Huazhong University of Science & Technology (HUST), Wuhan. The Competence Center is financed by the Kuehne-Foundation, Switzerland. This set-up clearly emphasizes the international nature of the work that was conducted here.

This Sino-German research cooperation has been fruitful from the very beginning, when we started the project in January 2005 with our first meetings in Wuhan. The following design of a questionnaire and the comprehensive analysis of the results were characterised by mutual collaboration and open exchange of ideas in a friendly atmosphere.

The report is intended to fulfil both the practitioners' needs for their daily work in globalised logistics networks, and the academia's scientific interests in the phenomenon of international logistics. We hope this survey contributes to a further understanding of the issue and helps those who create tomorrow's global logistics systems and the conditions logisticians will work in.

We sincerely thank all companies, logistics managers, and scientists who supported us during the work for this survey – either with interviews or by completing our questionnaires. We really appreciate your contribution.

With best regards from Berlin and Wuhan,

Prof. Dr.-Ing. Frank Straube
Head of the Logistics Department
Berlin University of Technology

Prof. Shihua Ma
Vice Dean, College of Management
Huazhong University (HUST), Wuhan

Dipl.-Ing. oec. Michael Bohn
Research Associate, Logistics Department
Berlin University of Technology

Background and Executive Summary

Over the past decades the world economy has reached an unprecedented level of global integration. As markets are being liberalised and trade barriers continuously being removed, companies are in an ongoing process of internationalisation to tap growth potentials in new regional markets and leverage cost advantages of locations with specific factor endowments and cost conditions. As a result, we see companies with internationally dispersed operations, embedded in global value chains and serving customers worldwide.

Background and scope. For the internationalisation of business activities, Global Logistics Systems play a significant role. They bridge the gap between cost efficiency in the global value chain and the required customer service. This clearly emphasizes that, nowadays, logistics is one of the most decisive success factors for companies in global competition.

The motivation of this survey is to review companies' internationalisation procedures from a logistics perspective. What are the challenges of setting up international logistics networks? What are the problems logistics managers must face when entering specific markets? How do managers cope with these challenges to ensure that internationalisation projects will be successful?

The research has been undertaken in China and Germany, surveying each country's companies when going global. In order to obtain a realistic picture, various aspects have been analysed according to the most important target countries and regions. These were identified as China, India, Russia, North and South America, and Western and Eastern Europe. However, the scope of this research is not a specific market analysis. The focus rather lies on patterns in internationalisation behaviour and logistics strategies for foreign market entry.

Executive Summary. This survey adopts a holistic view of the collective experience of Chinese and German companies in regard to how they organize and employ logistics and logistics strategies in the internationalisation of their companies, and the challenges they face herein. Thereby it becomes clear that while internationalisation is not new, and the overall environment for setting up shop in foreign markets is probably better than ever before, it remains a challenging task.

As the basis for all further analysis, the survey starts with the corporate objectives that drive companies to set up offshore business entities. It shows that companies follow a complex set of objectives that are mostly composed of two major issues: to benefit from specific cost advantages and to leverage growth potentials of the respective foreign market at the same time. Hence, logistics systems have to be implemented to integrate the new facility into the global network as well as it has to cater the local demand.

It can be understood that on the one hand the advancements in logistics are enabling internationalisation, and on the other, internationalisation is steadily challenging the logistics discipline with its heightened requirements in terms of time, quality, and cost. Thereby, it has to be distinguished between country specific and transnational issues. The most important local challenges when entering new markets are infrastructure (especially in emerging markets), security issues, and the cultural distance between the domestic country of a firm and the respective target market. Intercultural management is significantly gaining importance for logistics managers. Transnational issues arise from the integration of local logistics systems to form a global network. Most important is that companies are aware that situations in different countries are diverse and constantly changing. Hence, their logistics strategy has to cope with complexity and uncertainty. Regarding international trade procedures, it becomes clear that the main problem is not only the direct costs of tariffs and duties, but the lack of reliability in clearing, which may cause unexpected hold-ups.

The analysis of the internationalisation process forms the core of the survey. A simplified sequential process model was developed from the initial strategic idea to set up shop in a specific market until the final implementation of the new logistics system and its integration into the global network. Chinese companies finish this process within 14 months, Germans take a little longer with approximately 18 months, indicating the severe time pressure under which foreign market entries are carried out.

Improvements are necessary in the consideration of logistics aspects and the involvement of logistics managers, especially in the early phases of the foreign market entry, when strategic decisions have to be made. The most successful companies integrate logistics issues from the very beginning. Thereby, they do not only save time and money, but achieve competitive advantage as they meet the service expectations of their customers better.

Chinese and German companies follow different sets of objectives when establishing logistics systems in new markets; while the Chinese focus on a short and cost efficient set-up phase, Germans rather look at the characteristics of the final system to make sure that operations will be reliable, and time and cost efficient. Therefore, German companies accept a longer planning and implementation phase.

Three key components have been identified that have to be part of any logistics approach for setting up a offshore business entity:

Firstly, logistics managers have to ensure a high degree of flexibility and agility in their system to make sure it will meet the requirements of unknown future requirements. Basic levels of flexibility are generally incorporated among the reviewed companies, but it still falls behind cost and time considerations. A sound logistics system should provide companies with the ability to make swift adjustments in their supply chain in terms of where they manufacture, source, and sell. Surprisingly, few companies include an exit option in their logistics strategy, despite a growing number of famous market retreats.

Secondly, in order to deal with the heterogeneity and complexity of international networks, logisticians need to balance globally standardised procedures with local adaptations. The benefits from global process standardisation are widely recognised, yet not fully leveraged. Companies still tend to adapt their systems to a large extent to local requirements.

The third component is the strategic collaboration with logistics service providers (LSPs). Chinese companies traditionally outsource less than western companies. But in the context of foreign market entry the collaboration with LSPs can significantly facilitate to set up logistics operations. Hence, internationalisation leads to higher outsourcing levels and a broader range of outsourced activities at all reviewed companies.

To bridge the growing geographic and cultural distances in today's global supply chains, it takes a holistic logistics management approach that considers a multitude of aspects, ranging from the regional differences in lifestyle, working behaviour, technological skills

and equipment, infrastructure, and the supply of professional logistics services to the
subtleties of international trade. The most
important success factor for setting up shop
in foreign markets is to be aware of this diver-
sity and complexity, and to make sure that the
people involved are sufficiently trained and
empowered to get the job done efficiently.

Research Objectives and Methodology

This report is the result of a comprehensive research project on the Internationalisation of Logistics Systems conducted by the Competence Center for International Logistics Networks at the Berlin University of Technology (TUB), Germany, in close cooperation with the Supply Chain & Logistics Institute at the Huazhong University of Science & Technology (HUST) in Wuhan, China.

Research objectives. Whenever companies internationalise their corporate activities, the respective logistics system has to be expanded consequently. This might be the case, when new facilities are set up in Low Cost Countries or new regional markets are to be penetrated by opening sales offices or outlets. The resulting challenge for logistics managers is to design, implement, and monitor the adequate logistics networks for handling the corresponding flow of goods and information.

The objective of this survey is to develop a clear understanding of how logistics managers ultimately cope with those challenges type of problems they face herein – both from inside and from outside the organisation. It is looked at the complete process of setting up business activities in overseas markets, from the initial strategic decision to enter a specific region, until the final implementation of the local logistics system, and its final integration in the global network. Thereby, strategies and methodologies applied by logistics managers are analysed and compared between Chinese and German managers in order to identify differences and commonalities in their approach. As logistics nowadays contributes significantly to a company's effectiveness, it will be examined as to whether specific concepts are more likely to succeed than others and what the relevant success factors are.

This research is limited to companies who do not see logistics as part of their core competence, i.e. manufacturers and retailers. Having completely different business models employed and following different goals with their internationalisation, Logistics Service Providers (LSP) are not reviewed here.

Research methodology. The research project was conducted with a combination of qualitative and quantitative methods. Based on a comprehensive literature review of the Internationalisation of Logistics Systems accompanied by desk-top research on the

Fig. 1 — Research approach

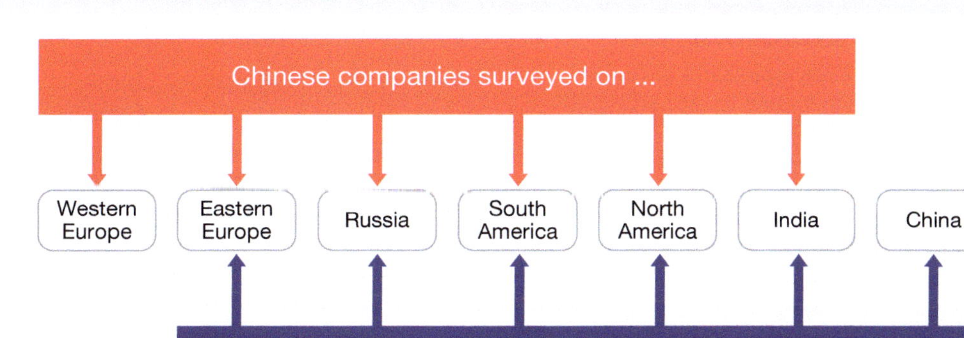

Fig. 2 — Sample representation

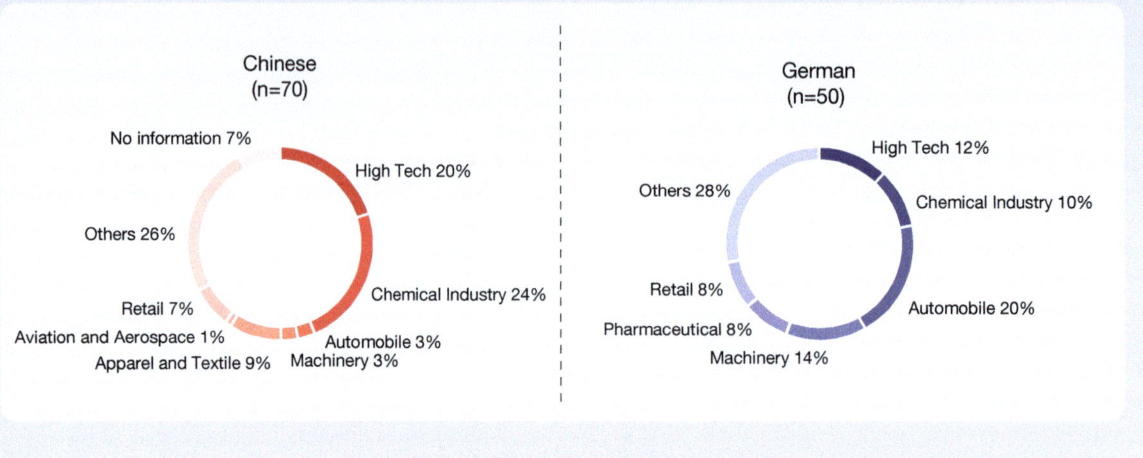

latest developments in the field, a general design of the research was developed. To get a broad and valid picture of internationalisation patterns along with the strategies and methodologies companies apply herein, it was concluded to base the investigation on a questionnaire based approach. The questionnaire was aimed at logistics managers from Chinese and German manufacturers and retailers, with global activities or with the intention to internationalise their activities. It was decided to review their global expansion towards a limited set of target regions and countries, which were identified to be the most important: North and South America, Eastern and Western Europe, India, Russia, and China (Fig. 1, comp. Chapter 1 for more details on the selection of these target regions.) The concept of the questionnaire resulted from the ideas and thoughts generated during the literature review, and was tested and improved by a limited number of in-depth interviews with selected practitioners in China and Germany.

The questionnaire was then distributed in German, English and Chinese among Chinese and German companies, with the option to answer either paper based or online. The poll started in June 2006 and the collection ended in January 2007. In total, 1,000 copies were sent out in Germany with a return of 50 valid replies. In China 900 copies were sent out and 70 valid replies were received. Some 20 % of the respondents answered online. Fig. 2 depicts the distribution of industries in the sample.

The following analysis of the resulting Chinese-German data set was undertaken jointly by the two project partners employing common statistical methods for empirical research. The evaluation method for significance assessment is based on the creation of the arithmetic mean of a Likert Scale. Simultaneously, discussions of the results with logistics practitioners enriched the interpretation of the numbers.

Along with the results of the poll, this survey provides additional information from various sources for the selected target regions and specific internationalisation topics.

Research Report

List of Excursions

List of Figures

List of Tables

List of Abbreviations

3PL	Third Party Logistics
APS	Advanced Planning Systems
Benelux	Belgium, Netherlands, Luxembourg
BRIC	Brazil, Russia, India, China
CAGR	Compound Annual Growth Rate
EDI	Electronic Data Interchange
ERP	Enterprise Resource Planning
EU	European Union
FBU	Fully Built Up
FCL	Full Container Load
FDI	Foreign Direct Investment
GDP	Gross Domestic Product
ICD	Inland Container Depots
ID	Identification
IT	Information Technology
KPI	Key Performance Indicator
LSP	Logistics Service Provider
M&A	Mergers and Acquisitions
MERCOSUR	Mercado Común del Sur (English: Southern Common Market)
NAFTA	North American Free Trade Agreement
R&D	Research & Development
RFID	Radio-frequency Identification
RMB	Renminbi: official currency of the People's Republic of China
SARS	Severe Acute Respiratory Syndrome
SCM	Supply Chain Management
SMEs	Small and Medium Sized Enterprises
SUV	Sport Utility Vehicle
TEU	Twenty-foot Equivalent Unit
UNCTAD	United Nations Conference on Trade and Development
WTO	World Trade Organisation

Internationalisation of Logistics Systems
— How Chinese and German companies enter foreign markets

1 Introduction—Internationalisation of Business Activities

The conditions for employment and execution of cross-border economic activities have improved significantly during the last century, and have lead to an unprecedented amount of globally exchanged goods and globally dispersed value chains.

Integration of the world economy. A comparison between the world GDP development and the world merchandise exports development as depicted in Fig. 3 and Tab. 1 demonstrates evidence of this strengthening interconnectedness. The liberalisation of markets, decreasing transportation costs, and huge steps forward in information and communication technologies not only lead to an increase of international trade in the most basic sense of products being distributed abroad, but moreover, lead to globally interconnected production systems. In this environment of increasing global integration of dislocated business activities, it is logistics that performs as the main enabler. By designing, managing, and monitoring the world's flows of goods and information, logistics is justifiably regarded as the backbone of the globalisation phenomenon.

Changes in the global market place. Although the triad of North America, Western Europe, and Japan concentrate nearly half of the world's GDP and are home to the lion's share of the Fortune Global 500 (see Fig. 6), current economic data foreshadows that the old triad-Era is becoming outdated. Rapidly emerging markets, sometimes referred to as BRIC (Brazil, Russia, India, China) have been showing impressive high growth rates and, considering the significant disparities between their share of the world's population and their share of world GDP (see Fig. 3), they have great potential for further persistent and sustainable growth. In addition, economies such as the Chinese are becoming more advanced and technologically competitive to their triad counterparts. They begin to move away from merely providing low-cost labour and bring forward capable and ambitious players into the international market place, altering the traditional roles between the developed and the developing world. Internationalisation of businesses is happening in both directions: from the developed into the developing economies and vice versa.

Most important targets for foreign market entry. According to their economic relevance for the world economy, the seven most important target regions and countries for internationalisation activities were identified for this survey: North and South America, Western and Eastern Europe, Russia, India, and China.

Due to their high current significance and positive future prospects, the countries China, India, and Russia are analysed as single countries. They show the greatest economic potential - although their combined GDP is currently low at only 8 % of the world's total, growth rates have far outpaced those of the great economic powerhouses over the last decade (see Fig. 4). China, for example, has experienced double digit growth for almost 30 years consecutively. The Indian average GDP growth has doubled over the last decade.

To keep the complexity of the survey at an adequate level, other countries are subsumed in the four regions of North and South America, Western and Eastern Europe. Although there is a certain heterogeneity within these regions concerning level of development (political situation and infrastructure etc.), each generally understands itself as a region in the economical, political (e.g. "EU", "MERCOSUR") and cultural sense, and is perceived as such on a global level. Western Europe and North America represent the group of devel-

Fig. 3 — Development of world exports and GDP, comparision of shares in world GDP and population

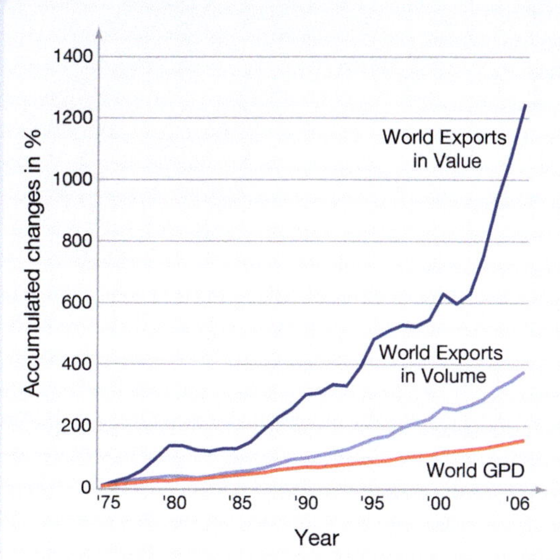

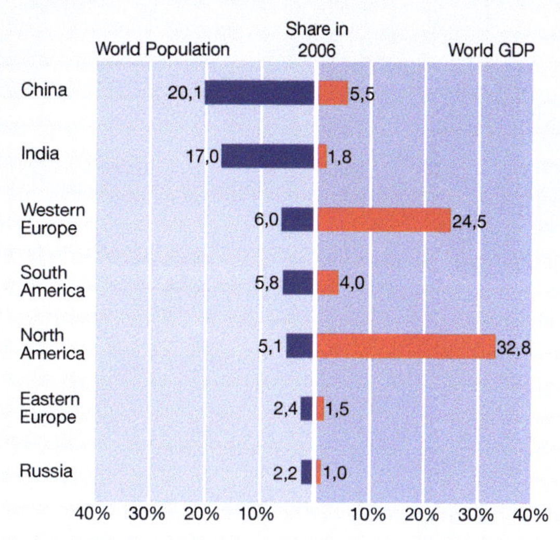

Source: ERS International Macroeconomic Data Set; Census Bureau of the U.S. Department of Commerce; World Bank World Development Indicators; World Trade Organization, Statistics Database; own calculations

Fig. 4 — Development of GDP over time, comparison of growth and wealth

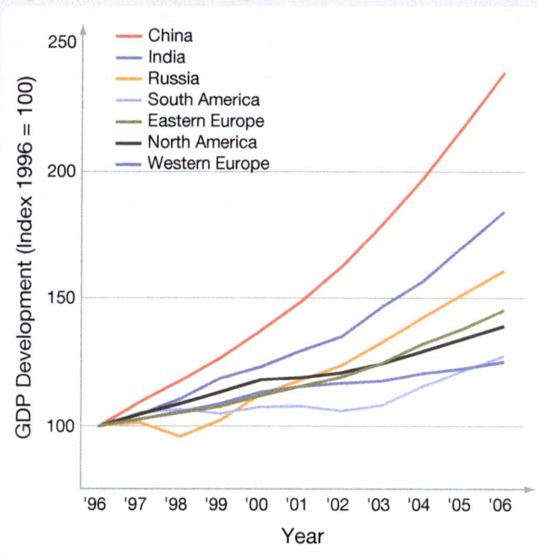

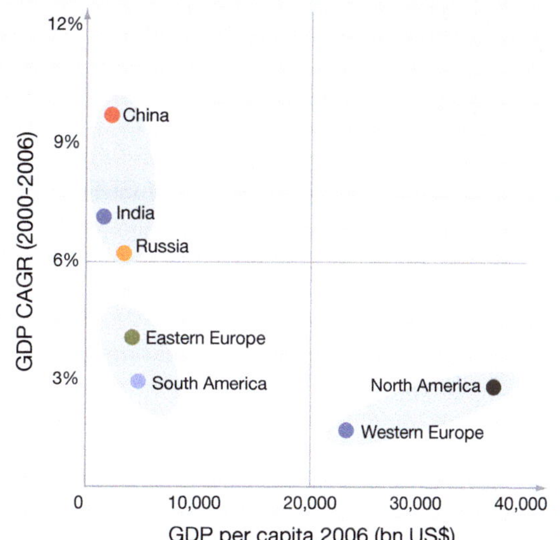

Source: ERS International Macroeconomic Data Set; Census Bureau of the U.S. Department of Commerce; own calculations

Tab. 1 — Key figures on the target regions and countries in the survey

	GDP 2006 (US$/capita)	GDP CAGR 2000-'06 (%)	FDI Inflow 2005 (m US$)	FDI Inflow CAGR 2000-'05 (%)	Exports of goods and services 2005 (bn US$)	Export CAGR 2000-'05 (%)	Imports of goods and services 2005 (bn US$)	Import CAGR 2000-'05 (%)
Western Europe	23,480.35	1.7	397,066.49	-12.64	4,549.04	8.5	4,408.96	8.5
Eastern Europe	3,549.05	4.4	50,651.20	6.68	478.86	18.7	505.21	19.3
North America	37,387.89	2.8	132,264.90	-12.99	1,551.17	2.1	2,134.13	4.9
South America	4,008.56	2.9	44,697.36	-7.66	354.53	12.7	268.60	7.0
Russia	2,608.64	6.1	14,599.61	5.53	268.32	18.6	164.62	21.4
India	612.24	6.9	6,598.00	3.96	165.49	21.1	194.80	23.8
China	1,579.07	9.8	109,073.69	4.78	836.89	24.5	712.09	23.2

Source: UNCTAD; ERS International Macroeconomic Data Set; Census Bureau of the U.S. Department of Commerce; World Bank World Development Indicators; World Trade Organization, Statistics Database; own calculations

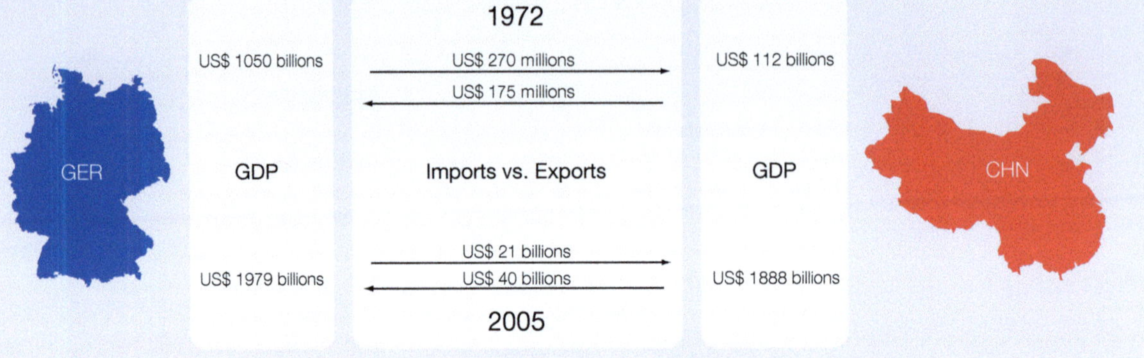

Fig. 5 — Economic ties between China and Germany

Source: ERS International Macroeconomic Data Set; Census Bureau of the U.S. Department of Commerce; World Bank World Development Indicators; World Trade Organization, Statistics Database; own calculations

oped economies, with comparatively high incomes per capita at low growth rates but with a huge share in global FDI inflows and outflows. Furthermore, these economies account for the largest number of global players in the marketplace (see Fig. 6 and Fig. 7).

The economies of South America just as Eastern Europe are also referred to as emerging or developing markets. The level of GDP per capita remains far below the developed economies and growth rates fall behind those of China, India, and Russia. Their position in the global market shows very positive future prospects as indicated by the shares of global FDI inflows.

With these countries and regions as the subject of discussion, this survey provides a comprehensive overview of the main targets of foreign market entry, while keeping a straightforward number of datasets. (For detailed information on the respective regions and countries, see individual excursions.)

How Chinese and German companies enter new markets. The survey looks at logistics strategies for global expansion of companies from China and Germany. These two countries are key players in the world economy with specifically high involvement in foreign trade, and both countries significantly benefit from globalisation (see Fig. 5). Furthermore, the Chinese and the German econo-

mies share close ties with each other: Germany is China's largest export market in Europe and receives the largest share of China's FDI to Europe. China is Germany's second largest export market in the world and accounts for the major share of German FDI to Asia.

German global players represent those companies that are based in a developed market and have gained experience from internationalisation for decades. Chinese companies, on the other hand, are representing those based in emerging markets. Having benefited mostly from the low cost labour endowment at home and being successful in domestic competition, they are now heading off for new markets in both developed and developing countries. This new breed of Chinese companies has the potential to cause major changes in global competition.

With a sample of 120 companies whose logistics managers gave details on their logistics strategies for going global, this survey gives deep insights into a variety of topics related to foreign market entry. Similarities and differences between the German and the Chinese approach are determined, and facts and figures are provided for a thorough understanding of the respective target regions.

Fig. 6   —   Distribution of Fortune Global 500 companies in 2006

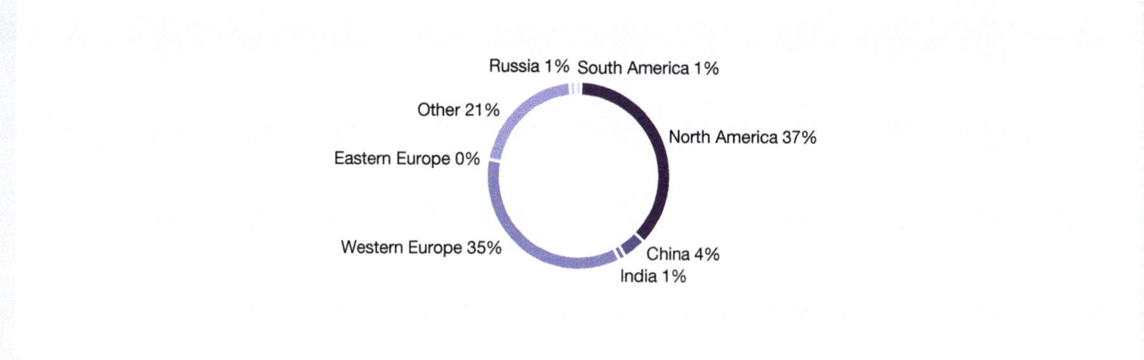

Source: 2006 Fortune Global 500

Fig. 7   —   FDI inflows per country (2005) in million US$

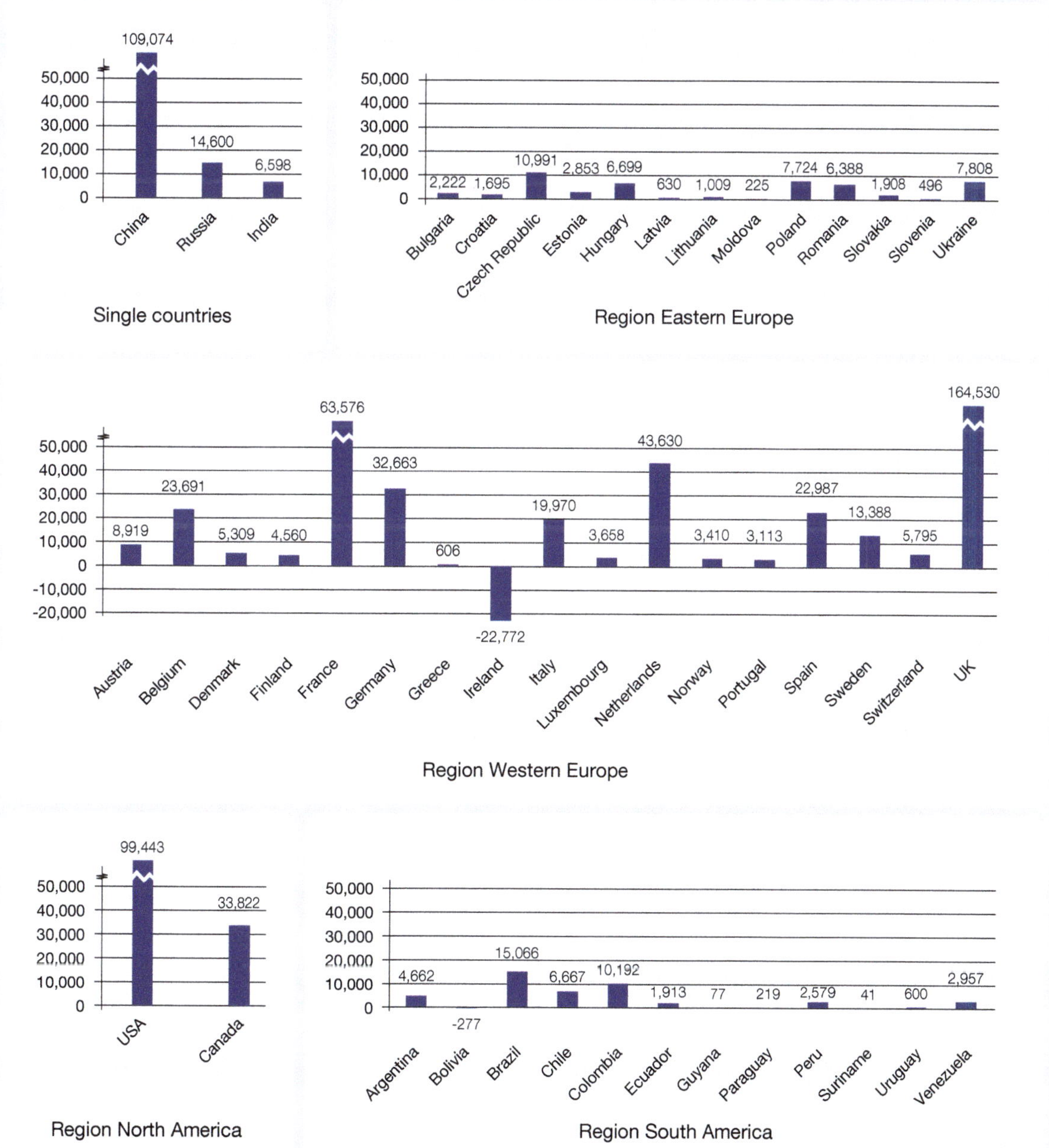

Source: UNCTAD databases

Insight — China

The Chinese economy has grown rapidly since the economic reform process began in 1978 and recently surpassed the United Kingdom to become the world's fourth-largest economy. Over the past 25 years, the mainland economy recorded an average real growth rate of more than 9.5 % per annum, making it the new world record holder among major economies. From only 4 % of global output in 2000, China should account for 11 % of world GDP by 2025.

In 2005 China is already the world's third largest trading nation behind only the US and Germany, with trade expanding nearly 29 % annually on average between 2001 and 2005, and seems destined to become the largest some time in the second decade of the 21st century. As an economy where external trade is equivalent to 72 % of the GDP, China stands to benefit from further rounds of trade liberalisation. The inflow of international capital has deepened China's integration with the global economy, attracting nearly US$ 230 bn between 2002 and 2005.

China joined the WTO in 2001 after 15 years of negotiations and declared the step as a strategic decision in the process of an economic globalisation. The public sector became less important. Enterprises without competitive advantages are getting eliminated from the market, whereas the private sector benefited from the WTO participation. First of all, private companies enjoy having the same opportunities as state-owned enterprises at present. Secondly, due to the commercial partnerships private companies created a quick access to foreign know-how, capital, and technology.

In the six years since the WTO accession, China has changed in many positive ways. The ideas of market economy, trade, and investment liberalisations have been integrated into popular thinking. WTO core concepts including transparency, accountable governance, and national treatment are now widely accepted by the Chinese public. China's WTO entry has also changed the global economic landscape. The country has become a major global manufacturing centre and the engine of the global economy. Global sourcing is focussing heavily on China nowadays.

The labour force consists of 745 million people, that is separated into 3 categories: agriculture (50 %), industry (22 %) and service (28 %). However, the unemployment rate in China is extremely high. In urban areas it reaches 10 %, whereas in rural areas it is even higher. According to Premier Wen Jiabao, the gap between the cities and the hinterland is escalating. Tensions such as rural economic stagnation, disintegrating education, unacceptable healthcare system, and high unemployment are growing more acute.

Hence, global consumer goods companies initially focused their efforts mostly on China's three largest cities, Beijing, Shanghai, and Guangzhou. These areas are home to some 29 million people and represent China's most affluent and sophisticated market, accounting for 13 % of the country's disposable income. But relatively few global companies are taking note of the roughly 12,000 smaller cities and towns that dot the Chinese landscape. The total household income in these cities and towns is already about 50 % higher than that of the so-called first- and second-tier cities combined. The number of households in smaller cities and towns with incomes greater than 35,000 RMB a year is expected to grow annually by 7.6 million, or 7 %, over the next two decades, compared with average annual growth in cities generally of 6.6 million, or 5 %. These smaller communities form the gateway between urban and rural China. Altogether about 400 million people - around a third of China's population - live in these regions.

As the world's third largest trading nation, China's logistics industry is gaining in importance. The government's 10th and 11th five-year plans stress the development of the logistics industry so as to facilitate and support continued economic growth. Currently, transportation costs constitute nearly 55 % of total logistics costs. Hence, China's government took the initiative in reducing this expense. In order to achieve improvements, it invested Rmb775bn in 2005 in infrastructure to address bottlenecks in the transportation network, an increase of 23 % over the previous year.

Much investment will be channelled into highway construction that grew at 6.7 % annually from 2001-05 to improve inland transportation. In addition, China is expanding its sea port capacities accomodate growing demand. Shenzhen and Shanghai, for instance, are ranked among the world's top four container ports regarding container traffic in TEU in 2005 behind Singapore and Hong Kong.

Airports in China have grown at a fast pace. More precisely, seven out of China's top ten airports have experienced two-digit growth rates over the last two decades. China's air freight industry, for instance, grew by 11 % in 2005 and is estimated to increase until 2010 by more than 50 %. In fact, China's transportation network is one of the largest in the world. Even in less developed regions such as the southwest, China will double its airports from 24 to 48 by 2010.

However, there is also a downside of China's growth. China's development relies strongly on exports which places the economy in a position vulnerable to external fluctuations. The market for logistics services is vigorously fragmented and underdeveloped. Experts estimate a total amount of up to 700,000 logistics companies in China. Not even one is able to possess 2 % of the market share. Due to these conditions 20 % of the GPD consists of expenses for logistics services, twice as much as in the US. 40 % of products' total costs are logistics related. China's exports to the U.S., for example, are significantly higher than imports vice versa. Consequently, the necessary return of empty containers to the Chinese mainland influences the price structure of shipping rates on these trade lanes.

Some experts warn that the real challenges in the future will be further market openings and the ability of China's leadership and economy to cope herewith. Others show more confidence and are looking forward to more market opportunities and expect positive progress, e.g. regarding the protection of intellectual property rights.

2   Strategic Objectives of Internationalisation

The logistics strategy is a subsystem of the business strategy and has to ensure the accomplishment of the corporate objectives. Consequently, in the first step of the survey, the strategic objectives that companies follow with their internationalisation activities are determined as the basis of all further analysis. As business objectives may vary for different regions it is determined what companies aim at when entering a particular target region or country.

Not all different objectives can be clearly distinguished as they may be closely related and have complex interdependencies. As illustrated in Fig. 8, it typically is a complete set of objectives that drives a firm to set up shop in a foreign region. The importance of each objective is displayed for German and Chinese companies and also for each specific region or country.

Objectives of foreign market entry. Basically, the two main areas of motivation to enter foreign markets are to tap either growth potentials or cost saving potentials.

Thereby, it is interesting that the growth aspect is valued as much more important than cost reduction. Going global strengthens the companies' market position as they leverage global opportunities. In opposition to frequent statements in public media, global cost pressures that make firms shift their production overseas - causing unemployment in developed economies - play a rather minor role in the current internationalisation behaviour of companies. In the first place, internationalisation means leveraging global expansion opportunities, i.e. it is to the very advantage of the company.

In the past, companies with a clear cost focus used to set up shop mostly in so-called "spe-cial economic zones" or "free trade zones" in low cost countries to leverage the beneficial cost situation for manufacturing. The merchandise was mainly meant for export. Typically these special zones have direct links to international sea ports and/or airports, and provide services needed for international cargo handling. Hence, logistics is facilitated and the resulting network comparatively simple. Today, as the set of objectives is in many cases more complex and many companies follow the double objective of cost reduction and growth at the same time, the new overseas entities have to cater both, the new market itself and also other regions with exports. Consecutively, complex logistics networks are necessary, with local supply and distribution structures being integrated into a global network.

Comparison of German and Chinese sets of objectives. Looking at the patterns of Chinese and German companies' objectives for their internationalisation, one main difference becomes evident: While German companies almost exclusively focus on growth and cost, the valuation of objectives by Chinese respondents is more equally distributed (see the two curves in Fig. 8).

As shown in Chapter 1 developed economies like the German suffer from little growth or stagnation, while the developing nations are booming in rapidly growing purchasing power. Along with the fact that these countries simultaneously offer significant cost advantages, to participate in these dynamic market development is of major importance for German companies.

Chinese companies on the other hand have a broader range of objectives. Besides increasing domestic competition and market opportunities outside China, foreign market entries

Fig. 8 — Main business objectives for internationalisation

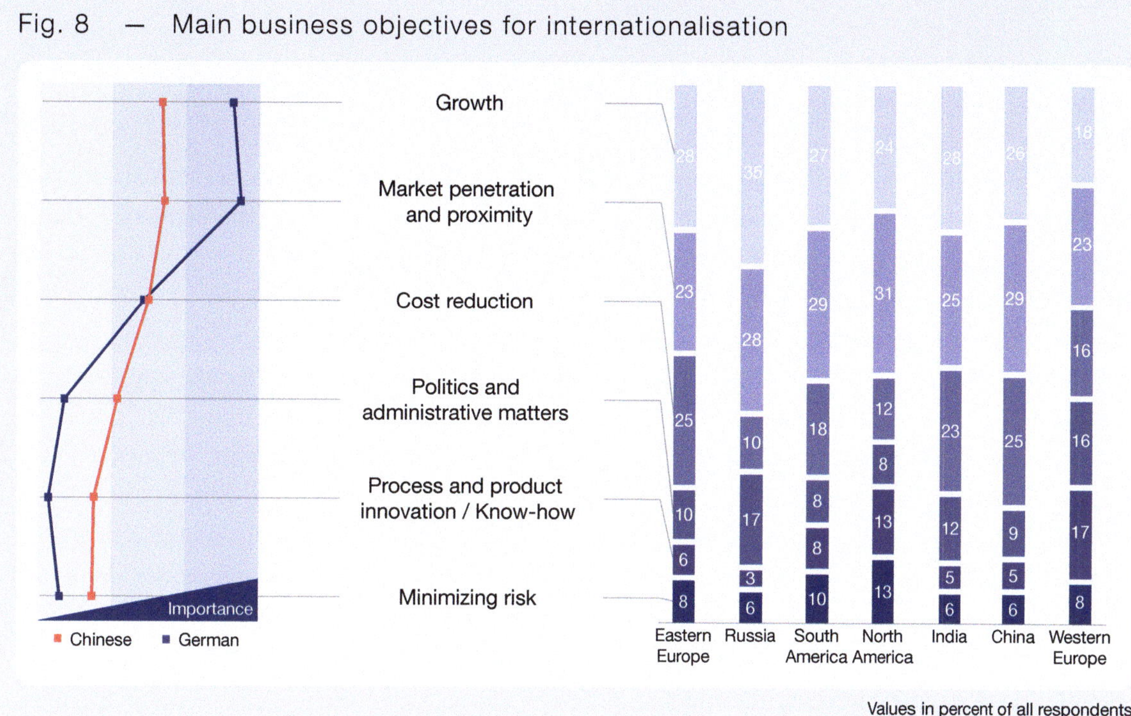

offer further opportunities in gaining access to innovation, know-how and specialised personnel or in balancing risk by a diversification of locations and markets.

Still, for Chinese, as for German companies, the growth aspect is most important, but with a distinct discrepancy: in many cases Chinese companies seek to leverage their cost advantages at home and enter foreign markets with a comparatively higher price level so they can achieve better margins. German companies, on the other hand, coming from a rather saturated market, seek to broaden their customer basis to realise economies of scale. As R&D expenditures per product have been constantly rising in the last years and product life cycles have been shortened, a broadened customer basis ensures an earlier break-even point and higher returns. Therefore, they even accept that the lower price level in developing markets lead to smaller margins.

At first sight, it might be surprising that Chinese respondents value cost reduction as similar important as their German peers, even though the overall cost situation in China seems to be much more favourable. But for specific products, manufacturing in developed countries may be more cost effective than in China. Especially for knowledge intensive products with high quality requirements the factor endowments of developed countries show specific advantages. The availability of well trained personnel, along with educated and sophisticated processes ensure lower quality costs in terms of waste, rework and reliability, and thereby outweigh the comparatively high labour costs. Thus, for both Chinese and German companies foreign market entries may enhance the overall cost structure.

Another objective to internationalise business activities is to reduce risk. This can be achieved by a deconcentration of manufacturing through the establishment of different locations, thereby securing reliable operations. It can also insure against currency fluctuations, and self-owned off-shore manufacturing can prevent losses of intellectual property, which may result from off-shore outsourcing to low cost countries.

Differences between regions. China and India attract foreign companies to set up shop by offering a combination of both cost saving potential, and a huge sales market with an ever increasing purchasing power. Hence companies establish structures to cater the local demand as well as for export.

The same is the case for the countries in Eastern Europe. While growth rates are comparatively lower than in China and India, the economic outlook is still very positive and the ongoing accession of more and more countries to the EU keep future expectations high. The labour costs for most parts of Eastern Europe remain low, but in certain areas and fields of industry wages catch up with the Western European level. Liberal tax regulations and factor costs create an overall favourable environment for market entries. A specific advantage of the Eastern European countries is their proximity to wealthy economies in the western part of the continent (EU 15, Switzerland and Norway), which makes them an attractive location for "Near-Shoring".

Currently, the main motivation to set up shop in South America is to cater to the domestic demand and participate in the positive market development of specific countries, e.g. Argentina or Brazil. Leveraging low cost manufacturing for export is of minor importance.

Russia attracts foreign market entries from manufacturers and retailers mainly with the increasing purchasing power of Russian customers. Unlike the emerging markets in East Asia and India, Russia is not a low cost location for manufacturing for the world market.

West European and North American countries are most interesting due to the high purchasing power of their customers. As mentioned above, cost reduction can only be achieved for specific products with higher levels of sophistication, as process and product innovations and know-how are of comparatively high importance to set up shop here.

In many cases, the final decision to locate an off-shore entity in a specific country depends strongly on politics and administrative matters. This includes subsidies, tax cuts, tariffs and non-tariff trade barriers, such as local content regulations and administrative procedures that hinder a company from catering a specific market with exports. These measures can have significant impact on the profitability of foreign market entries. Politics and administrative matters are perceived to be especially decisive in Russia, India, China, and Eastern and Western Europe.

3   Link Between Internationalisation and Logistics

There is a close relationship between internationalisation activities and logistics. On the one hand logistics is driving internationalisation as one of its key enablers. On the other hand internationalisation is challenging the logistics discipline with its unprecedented pace.

It is clear that the degree of globalisation we experience today would not have been possible without the advancements in global transport and communication systems. Transportation costs have been decreasing significantly during the last century, and so have communication costs. The costs for the most important international transport modes, sea and air freight, have fallen by approx. 65 % and about 88 % within the past 70 years. At the same time the transportation speed for goods and data has reached unforeseen levels, just like the capacity that today's global networks do offer.

For companies this has dramatically facilitated to set up off-shore facilities and the integration of overseas operations into highly complex value chains. Still, the design and management of these complex flows of goods and information remains a challenge for logistics managers worldwide.

The existence of global communication and advanced transportation systems frames an important prerequisite for global operations. But that does not imply that going global is an easy task. In order to successfully internationalise business activities, it takes a comprehensive management approach, incorporating logistics principles and methodologies.

3.1 Significance of Logistics for Corporate Success

In order to understand the significance of corporate logistics in the context of internationalisation it is necessary to begin with a look at the overall role that logistics plays in business today.

Over the past decades, the logistics discipline has been subject to constant change, as illustrated in Fig. 9.

Development of logistics over time. In the western economies during the '60s and '70s, classical logistics encompassed material and goods-related tasks and functions. Transportation, handling, warehousing, packing, and commissioning fell within this jurisdiction. At this time, great importance was already ac-

corded to the availability of raw materials and components. Logistics was perceived as only being able to marginally influence a company's success. It was not a discipline unto itself, but embedded in the corporate structure in a fragmented manner resulting in isolated logistical segments without proper connection.

In the '80s, a transformation from a function to flow oriented approach was recognized within logistics, where the goal was a configuration and optimisation of processes, by integrating and coordinating the previously fragmented business processes in procurement, production, and distribution.

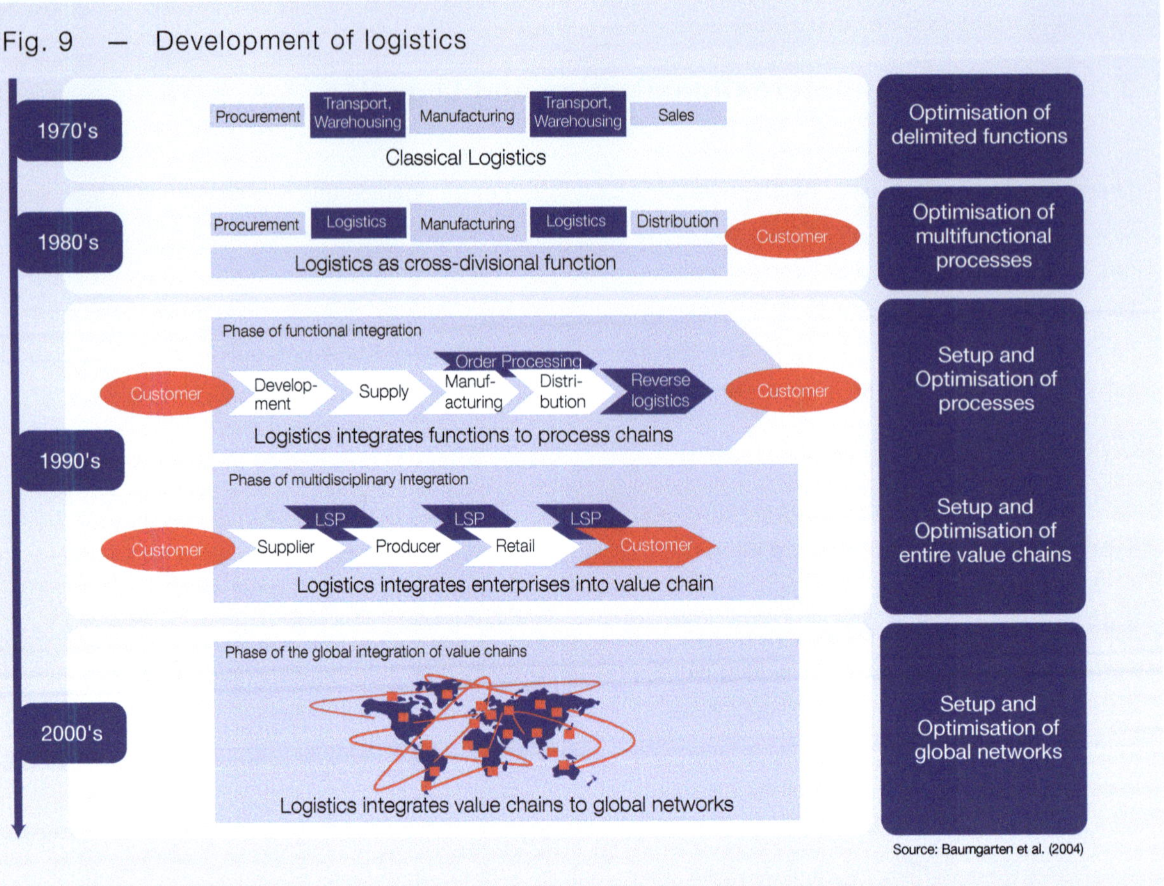

In the '90s, the concept of functional integration gained advances and incorporated the flow of goods and information. Through this holistic approach cross-functional optimization of the entire process chain could be achieved. The rapid advances in IT allowed the reduction of information deficits between and within the process chains.

Logistics has become a key function at strategic business levels as it has proven to achieve significant competitive advantage by catering the growing demands in speed and flexibility. Especially in the service driven and time based competition which is being observed in more and more markets.

Around the turn of the millennium, companies have begun to undertake the cross-functional and holistic coordination of business interfaces. Flows of goods and information are planned and controlled not only between functional entities but also across corporate borders aiming at a strong integration of suppliers and customers. This integration has been advanced through the concept of Supply Chain Management which is based on the wide spread of ERP systems, connected via EDI links. The upcoming Advanced Planning Systems (APS) are facilitating this development. This cross company integration is not limited to national boundaries — so, the current phase of the logistics evolution is characerised by the setup and optimisation of integrated global networks.

Status quo of corporate logistics. Today, very few companies worldwide have yet reached the level of integrated and globally optimised networks. Most enterprises from developed nations currently do adopt a process oriented and integrated logistics approach. They mostly have information exchange with their direct suppliers and customers and monitor certain key points of their supply chain both upstream and downstream. But the application of fully global integrated network approach is still in its infancy.

For average Chinese companies, like those from other rapidly developing countries, things are different. Most of these companies still see logistics as rather isolated functions, mainly transport and warehousing, that are undertaken independently at different departments of the enterprise, similarly to the logistics approach of the '70s and '80s in western economies.

Those Chinese companies that already run international operations or are about to internationalise their business activities, usually have a more advanced understanding of logistics. As exclusively these companies are included in this survey's sample, we discovered a different picture than expected for the average Chinese firm: research has shown that the upcoming global players from China value logistics at least as equally important as their western competitors. The most successful ones amongst them engage in sophisticated logistics planning when it comes to internationalisation. Due to the fact, that their market entries follow a Greenfield approach they have the chance to develop very country specific and strategically planned logistics systems.

So, it is worth mentioning that the upcoming Chinese global players may not be underestimated by looking at the average Chinese company, which is mostly focussing on domestic competition.

Today's role of logistics for corporate success. For 90 % of the surveyed companies logistics is crucial for competition, as customers require high levels of service in terms of logistics performance and reliability. Logistics service levels directly influence the purchase decision of the customer as he perceives the respective product in combination with its availability, lead time and other service aspects.

As Chinese companies mostly compete in commodity markets and in fields in which price is the most important sales criterion, their focus is slightly more on costs (Chinese: 92 %, German: 62 %), while German companies tend to compete in other areas, e.g. technological sophistication, quality, and service. Hence, German interviewees focus slightly more on service levels in competition than their Chinese peers (Chinese: 65 %, German: 79 %).

The critical aspect is that the fierce international competition in most markets has led to the situation that customers do expect high logistics service levels but are unwilling to pay extra for it, as indicated by more than 70 % of the managers.

Hence, as cost pressure and customer expectations are rising at the same time, effectiveness and efficiency in logistics systems become even more important for corporate competition.

Insight — India

The Republic of India is home to about 17 % of the world's population (1.11 bn people). Alongside China, India is regarded as number two in Asia's booming new markets and has become increasingly important to international business - foreign trade already makes up one third of the GDP while both imports and exports rose by 36 % in 2006 (April-December) to US$ 131.2 bn (imports) and US$ 89.5 bn (exports).

Over the past years, India's economy has experienced a growth of around 8 % per annum. Although India has been experiencing economic growth throughout the last 15 years, it is still considered a developing country in which around one quarter of its 1.11 bn people live below the poverty line. India's economy has experienced a forceful shift from the primary sector to the secondary and tertiary sectors in the past decade. In 2006/07 (1st April – 31st March), GDP was at US$ 923 bn, most of which was generated by the service and industrial sectors. In 2006, services – including retail, hotels, transportation, insurance, financial, social, and personal services – constitute 55 % of GDP. Although India is the world's second largest producer of food next to China, agriculture provides roughly 20% of GDP. At 1.5 % world market share, it only plays a subordinate role on a global scale. The agricultural sector provides the mainstay for the majority of the Indian population. Growth rates for the secondary and tertiary sectors have been at 10 % while the primary sector growth rate has decelerated from 6 % to 3 % between 2005 and 2006.

The country has become well-known for its IT expertise and is regarded as a location for offshore service outsourcing such as customer support. Besides services and IT, India also competes with world leaders in the production of pharmaceuticals, aerospace and biotechnology. Only recently, international companies such as global automotive manufacturers have discovered India as a production site. In order to reduce the dependency on offshore outsourcing, the Indian government is promoting the expansion of mass-production facilities.

Due to its socio cultural composition, the Republic of India is considered one of the most promising procurement markets. Factors contributing to this assessment are widespread proficiency of the English language, a stable governmental system, and broad endowment with well-educated engineers and technically apt workers. The Indian government is promoting foreign investments in manufacturing sites and infrastructure and has set up a legal framework that allows foreign companies to incorporate 100 % self-owned subsidiaries in certain industry sectors.

In India, economic activity is highly concentrated in major clusters in and around the largest cities. Delhi, Mumbai, Bangalore, Chennai, and Hyderabad together currently receive two-thirds of the country's FDI volume. The area around Mumbai on the western coast is home to the financial sector. This area provides nearly 40 % of GDP. In the south, IT-boom town Bangalore produces about 90 % of the nation's total software exports and is the location of aerospace and biotech companies. The automotive industry has built up capacity in the area around Chennai, former Madras on the southern Bay of Bengal. It is the designated home to Indian subsidiaries of international companies such as Ford, Renault, Mercedes-Benz, Hyundai, Honda, and BMW, as well as their subcontractors. Not far from Chennai, the biotechnology industry has settled in Hyderabad in the so-called "genome valley." Furthermore, the pharmaceutical industry – mainly focused on the production of generics – has settled around Ahmedabad in the northwest of the country.

As industries in these clusters continue to evolve and grow, there is a rising demand for qualified logistics services that are capable of providing qualified, complex supply chains. International logistics providers have made large investments in warehousing capacity. Local providers are yet not capable of providing nationwide services comparable to western standards.

The concentration of economic growth to these clusters causes challenges in several ways. Positive economic performance and higher incomes only affect a limited share of the Indian population. In rural areas, the situation has been largely unaffected by the growing economy. Thus, there is a general migration into cities, causing a rampant growth of slums in the areas surrounding the economic performers cited above. Disparities in the distribution of wealth may well be the cause for social unrest.

According to World Bank studies, major obstacles for investments continue to be bureaucratic hurdles, corruption, and deficient infrastructure.

Although infrastructure on the subcontinent is in a desperate state, until now national Indian investments in infrastructure have only been a fraction of the investments made by the most challenging neighbour China. While China invests US$ 200 bn in infrastructure yearly, India invests only US$ 28 bn.

On many routes, the railway system still operates at the technological level of around 1947 when India liberated itself from British colonialism. Yet, railways have provided the leading means of transportation for cargo and passengers until recently. As tariff policies intended to subsidize passenger transport burden freight transportation, there has been a noticeable shift to road transport of goods.

Roads are the main mode of transportation in India today. Although road density is comparable to that of the United States, road quality provides great difficulties. Less than 10 % of the country's road network is two-lane for each direction (only 1 % can be considered highways) and 40 % of India's villages are not accessible during poor weather conditions. As a result of chronic congestion and poor road quality, average truck speeds are around 30-40 km/h. Further delay is caused by the general lack of and obedience to traffic rules (such as lane driving) and a heterogeneous array of traffic participants. On almost all roads, including highways, truck drivers share the route with a heterogeneous array of road users such as livestock, harness teams, children, and cyclists. In an attempt to improve nationwide road-bound transport, the government is currently constructing the "Golden Quadrilateral," which, when finished, connects the cities of Delhi, Mumbai, Chennai, and Calcutta.

There are 12 major harbors in India and harbor capacities must be increased dramatically (doubled) in order to prevent a bottleneck for the nation's increasingly important foreign trade activities. At the moment, container handling times are around three or four times higher than global average.Until now, India has not managed to connect to international trade routes. Large freight volumes from India to Europe are still transferred to Singapore in feeder-transport and then forwarded to their final destination.

These infrastructural hardships, in combination with the governmental tax (service tax for outsourced logistics 12 %) and tariff framework make it difficult for 3PL providers to offer appealing quality services to customers. Yet, the Indian logistics market is expected to grow 18 % per year.

3.2 Logistics and the Success of Internationalisation Activities

Internationalisation of business activities means an increasing dislocation or geographic dispersion of the different steps in the value chain. It is clear that this leads to an increased significance of logistics in global networks compared to more locally concentrated operations, as logistics has to take care of the resulting flows of goods and information. Global supply chains show a higher degree of dependency on logistics performance and logistics costs account for a larger share of total costs. At the same time, the customers' expectations for logistics service and performance have increased (comp. Chapter 3.1). From these simple thoughts it is intuitive to conclude, that logistics frames an important backbone of any internationalisation activity.

Consecutively, it has to be determined to what extent companies and managers perceive the importance of logistics for internationalisation and whether it is sufficiently considered in respective projects dealing with foreign market entry.

Role of logistics for internationalisation. As depicted in Fig. 10 more than 80 % of the respondents see logistics as one success factor for internationalisation which has to perform in close interaction with other corporate functions. One third of the German interviewees perceive logistics as the most important enabler of going global, while twice as much Chinese managers are of this opinion.

About 15 % of all companies in the sample experienced situations in which specific market entries were not undertaken due to poor logistics conditions in the respective region. E.g. one company decided not to enter certain parts of South East Asia due to shortcomings in temperature controlled transport and warehousing.

Not even 20 % of the German logisticians indicate that logistics is not crucial for new market entries, among the Chinese it is even less.

It shows that Chinese companies focus even more on logistics in the global context than their German peers. This can be explained by several reasons. At first, Chinese companies experience a more challenging environment when they enter global competition by exporting goods. Due to insufficient domestic infrastructure conditions in large parts of China, a less developed market for logistics services and frequent shortages in shipping capacities going out of China, the logistics function played a key role at Chinese companies ever since, and especially for those engaged in going global. Consequently, managers of the upcoming Chinese global players have learned to focus more on logistics than their German peers whose exporting activities are embedded in a much more favourable environment.

Furthermore, Chinese companies compete mostly in low-end commodity markets with severe cost pressure. Hence, the cost consideration of logistics is much more crucial for them than for German companies who can afford higher logistics costs in their calculations.

Significance of logistics in different regions. Besides the overall recognition of logistics for the internationalisation, it can be observed that the interviewees value the importance of logistics for successful internationalisation varying slightly from region to region. Especially for China and India they see logistics performance as a very important success factor.

German companies state that especially in the emerging economies of China and India logis-

Fig. 10 — Significance of logistics for internationalisation

tics can make the difference between market success and failure. For Chinese companies especially developed markets like North America and Western Europe require specifically effective and efficient logistics to cater the comparatively high customer expectations in logistics service.

Summing it up, the logistics managers regard "good logistics" as one of the basic requirements for successful internationalisation and that it has to interact with other corporate functions. Logistics could be seen as a hygiene factor but with the potential to achieve competitive advantage.

As 86 % of the Chinese respondents and almost 40 % of the Germans frequently face severe problems due to an insufficient consideration of logistics in internationalisation projects, it becomes clear, that herein lies one of the key issues for foreign market entries. Top management recognition for the logistics network in budgeting, resource allocation, and strategic planning is most vital for global operations. This will be analysed in more depth in Chapter 5.

3.3 Impact of Internationalisation on Logistics

Just like logistics is enabling and driving internationalisation, internationalisation directly affects the characteristics of logistics systems.

Considering domestic and global logistics networks, one main difference is distance, which equates to transportation speed and dependency. Owing to trade lanes being much longer in a global setting, delivery and replenishment lead times are greater as well as unreliability of demand forecasts and uncertainty of actual product quality and quantity increase. Hence, internationalisation directly alters certain KPIs of logistics systems. In this chapter these changes are examined in combination with the corresponding strategies employed by businesses going global.

Increasing share of logistics in total cost. For several reasons, the internationalisation of the business activities leads to a rising share of logistics in total costs. Research shows that this increase is more severe for German companies than for Chinese companies. This is explained by the fact that even without internationalisation, logistics costs command a larger share of the total cost structure of Chinese products, due to shortcomings in logistics conditions in China.

Besides the increasing transportation efforts in global networks, internationalisation often causes product proliferation due to necessary local adjustments (frequently referred to as "localisation"). This causes rising logistics costs, as the additional number of items calls for more handling and administration efforts and less potential for economies of scale. This correlation is especially observed by Chinese managers. As German companies have increased the number of product derivates for years to cope with the challenges of individualised customer demand in Western economies, many Chinese companies only recently began to face the need for more product customisation. Accordingly about two thirds of Chinese respondents observe this challenge along with internationalisation, compared to only 45 % of the German managers.

Impact on inventory levels. More than two thirds of the German and Chinese companies face increasing inventory levels due to internationalisation as a result of extended transport distances.

Higher fluctuation in customer demand and reduced forecast accuracy also lead to rising inventory levels when companies expand globally. This, once again, is a challenge which is observed by more Chinese (about 70 %) than German companies (about 40 %), as these phenomena have already been widely recognised in the buyers markets of western economies since decades but are comparatively new to Chinese companies.

The application of related strategies to optimise inventory levels varies between German and Chinese companies:

- 63 % of the Chinese and 48 % of the German companies try to reduce inventory levels by centralisation of inventories.
- 74 % of the Chinese and 36 % of the German companies set up shop in high proximity to the market.
- 66 % of the Chinese and 44 % of the German companies are focussing on optimised planning and forecasting accuracy.

Transportation cost. 70 % of all respondants experience increasing transportation costs due to extended transportation distances, without significant differences between German and Chinese companies. In addition to the aspect of geographic distance, in international networks the use of expensive air transportation is generally rising as indicated by 63 % of the Chinese and 46 % of the German interviewees. This issue has to be taken into cost calculations from the beginning of any internationalisation project, especially when designing the business case. The general experience from all different branches of industry and retail shows, that in global networks air transport is to some extent inevitable. Even for goods of very low value density. There will always be some surprising incidents that can only be overcome by shipping merchandise by airplane, to keep schedules, satisfy key accounts, or simply to avoid stockouts of strategically important goods.

In order to reduce transportation costs in the global network, 75 % of the Chinese and 35 % of the Chinese companies try to benefit from optimised routing. One solution offered by various LSP is a combination of sea and air transportation, which would both minimise the costs whilst shortening the overall duration of the haulage. Especially between Asia and Europe, this concept is feasible and is gaining in importance.

Another approach, especially undertaken by small and medium sized enterprises (SME) is to form buying syndicates to get better cargo rates by pooling their freight.

Complexity of logistics network. A more subtle but still very challenging transformation of logistics systems when going global is the increase in complexity. 70 % of all respondents experience rising complexity due to the fact that more partners are involved in the network. 63 % of the Chinese and 45 % of the German companies stress the increasing number of involved LSP in global networks. One reason is that many companies tend to have specific LSP for each mode of transportation—and international transportation typically is intermodal by its nature, as the main haulage is either sea or air being accompanied by rail or road transportation. In addition, there is hardly any LSP at present who offers true global coverage at the required service levels and performance. Hence, companies tend to orchestrate different LSP for different regions and markets to cater their needs.

Insight — Russia

Since the dissolution of the Soviet Union in 1991, the Russian Federation has been in transition from a state-planned to a market-based economic system. This period of transition is still underway, hence mere reallocation of resources and reorganization of existing means of production can lead to huge efficiency gains.

Showing a persisting economic depression in the early post-Soviet years, the country plummeted into financial crisis in 1998. However, Russia recovered rapidly – mainly due to fortunate rising commodity market prices of its greatest export earners crude oil and gas. By 1999, GDP grew by 6.4 % and by 10 % in 2000. In 2006 the GDP increased by 6.7 % and this growth level may well continue. The unemployment rate has been decreasing over the last few years and is now at 7.2 % while inflation has decreased but remains high at 9 % in 2006.

With an area of roughly 17.1 million square kilometres, the Russian Federation (population: 142.8 Mio.) is the largest country in the world covering one ninth of the world's land mass. It is well endowed with natural resources and holds the world's largest natural gas reserves, the second largest coal reserves, and the eighth largest oil reserves. It is estimated that the oil and gas industry amount to approximately 20-25 % of its GDP. Thus, the Russian economy's direct dependency on volatile international commodity markets will continue to be a source of risk. The structure of exports is heavily dominated by fuels and raw materials generating a trade surplus of US$ 118 bn in 2005. Oil and gas account for 61 % while manufactured goods make only 10 % of all exports. Manufactured goods exports suffer from high export prices due to overvalued exchange rates.

Accumulated FDI remain low in comparison to other growing economies at US$ 52,518 million in 2003. Russian annual net inflow of FDI was at US$ 3,461 million only 6.5 % of its Chinese equivalent. Fixed capital investments are at a relatively low 18 %, most of which are concentrated on the dominating oil and gas industry. In its "Strategy for Future Social and Economic Development of the Russian Federation" (October 2003), the Ministry of Economic Development and Trade emphasizes the importance of diversification of Russia's economy away from natural-resource production in order to facilitate a more broad-based growth. The Russian government has distinguished transport and logistics as one opportunity of diversification. Russia is eager to establish itself as a hub for Asian-European transport.

Other key industries next to the oil and gas industry include fast moving consumer goods, high-tech/electronics, heavy, automotive, and the pharmaceutical industry.

Another threat to Russia's continuing development and growth lies within the high degree of concentration of wealth. The fruits of Russia's positive economic development since the crisis of 1998 have been unevenly distributed and disparities in regional economic development could well be an obstacle for sustained growth and reduction of poverty.

Although poverty still is a great challenge to the Russian Federation, the economic development and overall rising real incomes (10.7 % in 2006) have boosted domestic demand. Consumer spending is mostly focused on fulfilling pent-up demand and private investments are low.

The current political situation is marked by accumulation of power by the central authorities and a general authoritarian shift. Main points of concern include state control of media, lack of transparency, widespread nepotism, and corruption. However, the government initiated several reform measures prior to the upcoming WTO accession. The tough demands of WTO-membership require legislation pushing for more transparency and independent legal protection as well as a further opening of the Russian market and less bureaucracy. Such reforms include financial and capital market reforms, abolishing monopolies (e.g. telecommunication and railway sector), introduction of anti-monopoly laws, and simplified customs clearance. Even though several incidents in the past have shown worrying developments of state-influence on market players, foreign investors evaluate Russia's political situation as stable. The business situation is often described as connection- and favour-based business culture – the so-called "blat".

In addition to the natural challenges arising from Russia's mere size and long distances as well as adverse climate conditions, infrastructure such as telecommunications networks, health and transportation systems are in a state of decay. The government is addressing these issues by increasing budgets. One example of Russia's interest in improving its networks is the extension of the Russian-Chinese railway connection.

The Russian railway network provides a solid base for further development in commercial transport. It is the world's second largest railway system, covering 85,500 km of railway tracks. Most of the Russian main lines are owned by state-owned monopolist Russian Railway, which is Russia's second largest company in revenues next to Gazprom.

Rail freight is expected to account for around 40 % of total cargo shipments in Russia and for over 83% of cargo shipments excluding oil and gas. Railway passenger transportation represents 34 % of the transportation mix and is second largest next to automobile travel.

The most important current railway project is the extension of container transportation between Asia and Western Europe on the Trans-Siberian Railway. As president of Russian Railways Vladimir Yakunin noted in 2006, the main tasks facing Russia's railways today was to increase international transport on the Trans-Siberian and the Baikal-Amur Main Lines to 1 million TEU per annum. This project includes contracts with the German and the Chinese Railway as well as the construction of 18 container terminals. Russian Railways is planning to invest around US$ 4.3 bn in the Trans-Siberian from 2006 till 2008. Further projects include the reconnection of the Trans-Siberian to the Trans-Korean Railway and a continuing overhaul of the network.

Investments in transportation infrastructure have been low throughout the 1990s and have been increased only recently. In the short and middle term prospect, however, the desolate state of Russian infrastructure as a whole is not expected to improve significantly. Although infrastructure and bureaucracy are constraining factors, the logistics market in Russia is booming, struggling to keep up with logistics demand. In 2006, the market volume for logistics services in Russia was estimated to be US$ 120 bn. Around 6,000 companies are operating in the sector, most of which provide transportation services. As the Russian economy is proving solid growth, and foreign automotive manufacturers are transferring some of their production to Russia, there is a growing demand for sophisticated logistics services. Thus, international logistics service providers have recognized great opportunities in the Russian logistics market and are investing in their local infrastructure.

Fig. 11 — Transnational and country-specific challenges in global logistics

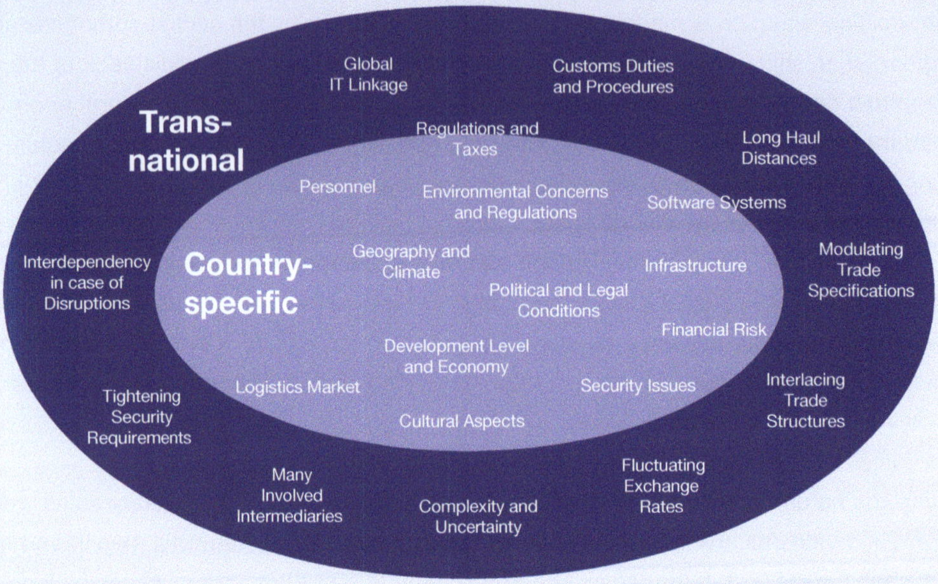

Source: according to Pfohl, H.-C. (2004)

4   Challenges and Difficulties in International Logistics

Heterogeneity is probably the most important challenge for international logistics activities. Each country and region has its specific elements of peculiarity, which lead to a great variety of domestic logistics systems which, combined have to be orchestrated in a global network. Hence, for designing and managing such a global logistics network, several categories of differences between the involved countries and regions need to be considered. Amongst others, these are mutual integration of commerce and trading structures, homogeneity of customer requirements concerning transport time and flexibility of delivery date, transportation infrastructure and administration as well as the geographical spread.

The procedures of trade and transportation are only partially standardised on a global level, but to a large extent they depend on national specifications.

Consequently, matching the global logistics strategy of a company with environmental heterogeneities includes a universal view of the logistics conditions both within involved countries and also transnationally, as illustrated in Fig. 11 above.

In addition, two aggravating circumstances have to be kept in mind: firstly, most of the countries differ not only from one another, but also internally — one might think of the differences between China's coastal areas in comparison to its hinterland. Secondly, many conditions are continuously changing, making global logistics very alterable and by no means static.

4.1 Country Specific Issues in Global Logistics

Country specific issues in global logistics include a variety of aspects. Some of them typically differ from country to country; others will be of great similarity between countries and regions.

Tab. 2 shows the most severe challenges for each particular region and country that logistics managers are currently facing there.

The differences in the evaluation of each region between Chinese and German companies clearly show two things: firstly, the severeness of the challenges in a specific region is a matter of perception that depends strongly on what situation the company is used to in its domestic market and also in relation to the overall problems in a specific region. I.e. in North America the overall conditions for logistics are comparatively good, hence soft aspects like intercultural matters are more likely to be perceived.

Secondly, the results show certain patterns that are related to the specific situation each society and economy is currently in. To name the most important ones:

Infrastructure is a typical issue for all the developing markets addressed here: Eastern Europe, Russia, South America, India, and China. It is clear, that shortcomings in infrastructure are especially problematic for further development and growth of the economy as a whole. From a company's perspective, the understanding of these shortcomings is very important for cost considerations — it is estimated that in developing countries, the share of logistics in total cost is about double, compared to developed countries. But in local competition it is of minor corporate importance, since infrastructure is equally good or bad for all players in the market.

Security issues of certain regions are by no means limited to logistics and have to be understood in a general context of society and economy. And from a logistics point of view it is of vital importance, since logistics handles all kinds of valuable goods, theft, robbery and damage will lead to poor reliability, increasing lead times, and high direct and indirect costs. Here, the regions of South America, Eastern Europe, and Russia are clearly under pressure. Various sources currently rank some nations of Eastern Europe as "most hazardous" in terms of transportation.

Intercultural issues and the level in which they are perceived is clearly related to the cultural distance between the home country and the new target market. As logisticians are considered the most important corporate link between the two, intercultural management is clearly gaining importance for their day-to-day business.

In the following, each country specific aspect is elaborated in more detail.

Personnel. The availability and skills of human resources clearly varies between countries. While unskilled labourers such as truck drivers and warehouse staff can be found around the globe, educated employees with technical and managerial logistics knowledge are scarce in many countries. This is especially true in emerging markets, where companies from any industry face a shortage of adequately trained personnel, both because of poor local education and shortages in supply of workers during rapid growth phases. Another important issue related to personnel is the willingness of expatriates to be appointed to leading positions in a specific market, as the attractiveness of different countries for upscale managers varies quite a lot. Usually less attractive locations call for higher

Tab. 2 — Country-specific challenges and problems

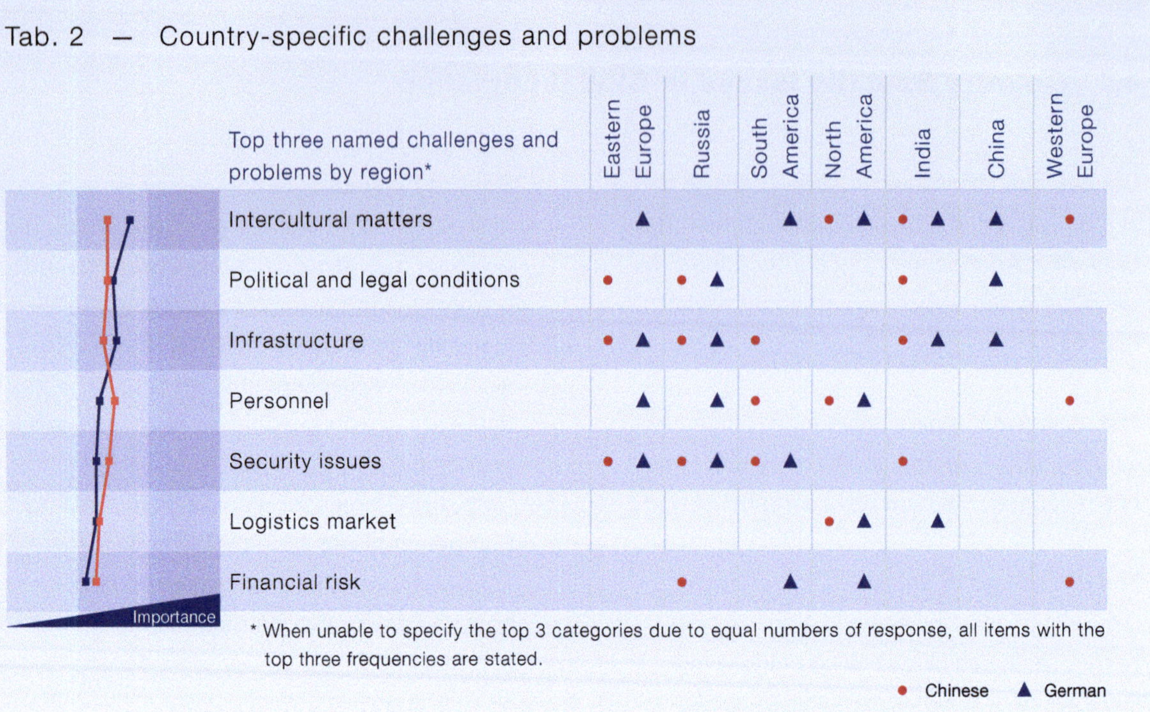

Top three named challenges and problems by region*	Eastern Europe	Russia	South America	North America	India	China	Western Europe
Intercultural matters	▲		▲	● ▲	● ▲	▲	●
Political and legal conditions	●	● ▲			●	▲	
Infrastructure	● ▲	● ▲	●		● ▲	▲	
Personnel	▲	▲	●	● ▲			●
Security issues	● ▲	● ▲	● ▲		●		
Logistics market					● ▲	▲	
Financial risk		●	▲	▲			●

* When unable to specify the top 3 categories due to equal numbers of response, all items with the top three frequencies are stated.

● Chinese   ▲ German

---

monetary incentives. Hence, it is important to foresee in early planning stages what efforts are necessary to train locals and to compensate overseas managers to live in the specific region.

Regarding expansions into economically further developed countries, the key issue logistics managers have to face is that competent workers are generally available, while unskilled labour, however, is relatively expensive in these countries.

Intercultural matters. At first sight, one might be surprised to find this topic related to logistics. But as the logistics discipline has broadened its scope in recent years, logistics managers today have to increasingly cope with aspects from various fields that were traditionally not considered logistics (comp. Chapter 3.1). Intercultural Management is one of these. Research has shown that knowledge in this subject is vital to today's logisticians, as differences in life habits, mentalities, languages, beliefs, values, customs, and environmental concerns are significant key challenges when interacting with people in global networks.

Looking at business practices, aspects of culture have to be considered from an organisational as well as from a customer point of view. Here, the term 'organisational' implies company internal procedures such as structuring of work and efficiencies in having a job done in each country. The customer point of view means the importance to understand the local needs and expectations of customers to assure the required service levels and appropriate treatment. The valuation of time for instance varies significantly between cultures, which might have great impact on service and goods movement. Punctuality may be viewed in some places as very important, while in others it is of minor significance.

In the intercultural context not only is communication and coordination made more difficult, but distinct local, country-dependent shifts in performance and motivation of logistics staff also exist. Moreover, these differences are constantly in flux because of cultural exchanges and assimilations of living standards and education within the dynamics of globalisation.

Consequently, culture has to be considered independently and distinctly when operating in different countries. Language is one of the most obvious differences between cultures and significantly influences the mutual understanding. The meanings of words often have to be seen in the general context in which they are used - as language is the common means of expression within a culture, it is inherently laden with meaning that might be obscure to the cultural outsider.

Another important aspect in global networks are religious practices, also procedures of prayers and sacrifices that have to be considered and treated with a maximum of respect. This has especially integrated in the planning and organisation of work, regarding working hours, holidays, places of worship etc.

Infrastructure. As the facilitation of goods movements is one of the fundamental factors governing the growth of a country's economic power, transportation infrastructures vary dramatically between nations at different levels of development.

In developed countries, the infrastructure is not only highly upgraded and superior, linking its industrial centres nationally, but it is also integrated to connect the country with its neighbouring countries. Recent challenges faced by developed nations are increasingly congested traffic lanes.

In developing nations, several reasons might have led to insufficient transportation networks and facilities. The rapid economic growth and, hence, the extension of domestic goods movements can overwhelm the existing infrastructures, which were often developed only for the transportation of basic goods. Furthermore, the focus on infrastructure development may initially have been driven by exports, thereby ensuring trade growth and the inflow of foreign exchange at first. This

could have hindered the building of a domestic system which companies rely on when they, as observed today, enter new markets with the objective to grow by catering local demand.

For example, outside the main economic centres in China, insufficient integration of warehouses and distribution facilities, as well as the lack of a suitable infrastructure are the main reasons for holding businesses back from outsourcing their operations to the western provinces, which would offer cheaper manufacturing and therefore reduce operating costs.

The quality of traffic routes depends not only on the surface conditions such as uneven and porous pavements, but also on the availability and density of networks. In some countries, the rate of network integration is very low, traffic links are simply missing and many highways and flyovers simply do not exist. Often, infrastructures are very fragmented domestically and include many regional differences between the level of quality of the networks and the level to which they are maintained. Moreover, other factors may lead to congestions, reduced speed and accidents, and therefore to increased transport times and associated costs. These factors include poor traffic control systems, missing traffic signs, inadequate traffic education and awareness, and slow moving vehicles such as tractors and cyclists on the roads. On Indian roads, for example, approximately 250 km can be travelled per day, while in developed nations over 600 km can be covered within the same time.

Logistics market. For decades, logistics outsourcing has been a very important method to lower logistics cost and to improve logistics quality at the same time (comp. Chapter 5.6). In the global context it is gaining even more importance and the range of services outsourced to logistics service providers for overseas operations is usually broader than it is at home.

Though, the quality and the range of logistics services offered in specific markets vary significantly and until today the global LSPs cannot offer their services in all countries and regions at the same standard. For example, China's warehousing sector was closed to foreign investments until the end of 2005. As a result of the lack of investment, China's low-end facilities, often with low ceilings, insufficient lighting, rare dock-levelling and inadequate security, cause high levels of product damage and stock deterioration. For instance, it is estimated that around 25 % of China's fruit and vegetable harvest is damaged every year by the inability to store it appropriately. In addition, the utilisation of load transfer points often differs between countries, some of which might bias the domestic movements of goods.

In India, for example, Inland Container Depots (ICD) are placed close to industrial centres and main metropoles to carry out customs clearance and distribution, helping to reduce congestion at the gateway ports. Consequently, import containers are transported to these depots duty free, primarily using railways. Hence, inland transportation networks have to be managed depending on infrastructure initiated by the Indian government.

As we will see in more detail in Chapter 5, a comprehensive market analysis regarding LSP services before entering a new market is of great importance and a clear strategy of how to organise logistics activities in collaboration with a capable LSP is regarded as a key success factor for the internationalisation of logistics networks.

Political and legal conditions. International logistics operations have to follow both global as well as national regulations. The survey shows that the political conditions in the emerging markets of Eastern Europe, Russia, India, and China are perceived to be specifically challenging.

Most emerging markets are characterised by institutional voids, leading to high importance of informal relations. The most important aspects to look at when setting up operations in new markets are: levels of corruption and nepotism, importance of informal connections to local authorities, arbitrary decisions from governmental authorities, security in jurisdiction and political stability.

Especially in German companies the importance of informal relations to authorities and even with local suppliers is sometimes underestimated. But as written contracts and laws are in many regions of less importance than in Western Europe, this aspect has to be taken into consideration from the beginning of any internationalisation activity.

In recent years, many countries have made huge efforts to facilitate and simplify regulations regarding the transportation of goods (e.g. liberalisation of logistics service markets), especially for import and export operations. Still, overregulation is perceived in many countries, in developed markets as well as in rapidly developing countries. Here, especially Russia has made significant improvements since 2004 with a reduction of customs regulations from some 3,500 down to 439.

Financial risk. As logisticians control significant assets along with warehousing facilities and inventory, financial risk management has gained considerable importance in recent years, not only in the context of internationalisation. With global operations, a company is exposed to different kinds of financial risk which may vary from country to country. Especially the experiences from financial crisis (e.g. Argentinean Crisis, Asian Crisis etc.) have shown the importance of the topic and logistics and supply chain managers are obliged to keep an eye on it when designing their networks, flows of goods, replenishment strategies and so forth. Currently logistics managers from both China and Europe see this issue to be most severe in Russia, and North and South America. For certain countries property rights remain a concern but most important is the volatility of currency exchange rates. With hedging strategies and optimised terms of payment and delivery, companies try to reduce those risks and to mitigate the respective consequences.

Security issues. Even though many companies mainly focus on increasing efficiency, security issues have become another major concern, especially for global logistics. The issue is of course not limited to national or regional boundaries. Still, as the survey shows, the valuation of the problem from logisticians' perspective does differ between the respective regions.

Security issues consider not only criminal acts such as theft, hold-up and fraud, but also increasingly politically motivated attacks, which might target trade lanes.

Today, criminals target not only high-value consumer goods, which have been typically most susceptible to theft, but are moving to other medium-value goods and different types of industrial cargo as well. The growth in global organised crime, which targets at large volumes of goods, suggests a strong reason for the increase of losses worldwide.

When one observes the domestic challenges of protecting cargo from theft and fraud, individual security standards of each country become evident. In certain parts of the world, poor economic conditions and organised crime cartels have led to increased threats to stored and transported goods. Whilst robberies are certainly a risk to the security of goods, hold-ups, in particular, incur additional risk to the safety of drivers and employees, and it is especially this added danger which is necessary for the implementation of avoidance measures. In certain countries, some LSPs organise convoys of trucks, employ shifts of drivers and appoint armed guards to accompany inland transportations. In response to the heightened risk of theft in Eastern Europe, some shippers choose to move cargo by sea from Germany to Russia rather than by land, which takes three days longer than by truck and is even more costly.

Insight — Western Europe

The term Western Europe was largely defined in times of the Cold War, but it is still commonly used even 17 years after its end in 1990. In the present survey, Western Europe includes Austria, Belgium, Denmark, Finland, France, Germany, Greece, Ireland, Italy, Luxembourg, the Netherlands, Norway, Portugal, Spain, Sweden, Switzerland, and the United Kingdom. With the exception of Norway and Switzerland, these countries also essentially make up the European Union (EU 15) before the 2004 enlargement, which saw many Eastern European countries joining the EU.

While this region's population only forms 6 % of the world's population, it commands for 24.5 % of the world's share in GDP. At US$ 397 bn, Western Europe has easily the largest FDI of all regions considered in this study, more than 2.5 times that of North America, which has the next largest FDI. Most of the countries in this region may be classified as developed nations. Western European economies are characterized by large GDP's per capita and technological advancements as evidenced by the fact that Western European companies also constitute for 35.4 % of all companies in the Fortune 500 list of 2006.

Due to the eastern enlargement of the EU, Germany has become more important from a logistics perspective. Geographically, Germany now is at the very heart of Europe providing access to many European markets. It's the continent's commercial hub, connecting North and South, East and West. As of 2006, Germany, characterized by a highly developed economy and a perceived high quality of life, has the largest economy in Europe and the third largest economy in the world, behind the United States of America and Japan, and is also the world leader in exports and world's second largest importer. Logistics is one of the key pillars of Germany's competitiveness as a business location, paving the way for added industrial value, the movement of goods and cooperation between companies. Behind the automotive industry and retail, logistics is the third largest sector in Germany, employing some 2.6 million people. The logistics sector also displays better than average growth. This is due not only to its geographic situation right in the heart of Europe, but to the top international position Germany has assumed in infrastructure and logistics technology in the view of many foreign investors.

The EU enlargement causes a relocation of logistics sites in aid of the eastern part of Europe. New markets are emerging with the potential for near-shoring facilities. That results in a changing logistics structure with the objective of a comprehensive Pan-European market place. Additionally, distribution centres were shifted from Benelux to Eastern Europe, to Poland for example, due to low wages comparatively to Western Europe. However, the supply of entire Europe with goods from Asia and North America is mostly transferred via Western Europe, especially by the use of harbours in Rotterdam, Hamburg, Antwerp, and Bremen – also known as the North-West-Range.

A secondary positive effect of the European Union is the harmonisation and coordination of Europe's transportation infrastructure. Western Europe's roads, railways, and airports compared to the world's average is highly developed and of great importance for logistics services.

The North-West-Range holds an outstanding position among the North Sea harbours concerning their size and their variety of function. They can be regarded as multi purpose harbours. Concerning total cargo handling Rotterdam is the largest European harbour, and it is also ranked 2nd in terms of cargo volume. Rotterdam, Hamburg, and Antwerp are also among the top 10 ports in terms of container traffic. The total of container handling in Europe amounts to 61 million TEU. Not only harbours of the North-West-Range have growth potential, also harbours on the Mediterranean grow by 9 % annually such as Gioia Tauro and Genoa (Italy), Algeciras, Valencia and Barcelona (Spain), and Piraeus (Greece). It must be pointed out that 4 % of the handled TEU in Western Europe is due to intra-regional transportation, whereas the main container harbours mostly operate as transit harbours as well as serving hinterland needs.

Western Europe is also home to some of the world's most important airports. The four major airports in Western Europe are Frankfurt (Germany), Paris (France), London (UK), and Amsterdam (The Netherlands), all of them situated geographically together. They unite 67 % of the regional freight traffic (5.4 million tons) and hold therefore a hub function for Europe with main focus on the international freight traffic. In Frankfurt, for instance, the Asian market covers with 771,000 t a share of 49 % of total air freight. In contrast, London focuses on trade with North America.

Overall, Western Europe benefits from a high diffusion of modern logistics concepts for the demanding industry and a comparatively high developed logistics market. By now, 24-hour transportation and delivery within Western Europe is already a common practice compared to Eastern Europe where it could not put into practice yet. Looking further, contract logistics will play a major role in an international context as well as globally acting logistics service providers. Still, outsourcing of logistics services in Western Europe is traditionally higher than in other parts of the world (e.g. China).

In Western Europe, technical and environmental standards are of high concern. The ISO 14001 environmental management standards, for instance, exist to help organisations minimize how their operations negatively affect the environment. Such ecological standards already have a high penetration at the logistics sector in Western Europe. In order to enforce sustainability in the field of logistics, for instance, one political approach is road pricing to support ecologically friendly railways. It is imaginable that these interferences will have long-term effects on the modal split for Western and Eastern Europe.

As the EU continues to enlarge, further cooperation between East and West is expected. As this cooperation deepens, all the more exciting is the prospect of bridging historical barriers brought about by political ideologies, as Europe progresses towards a united economic and political entity.

Fig. 12 — Main transnational issues in global logistic networks

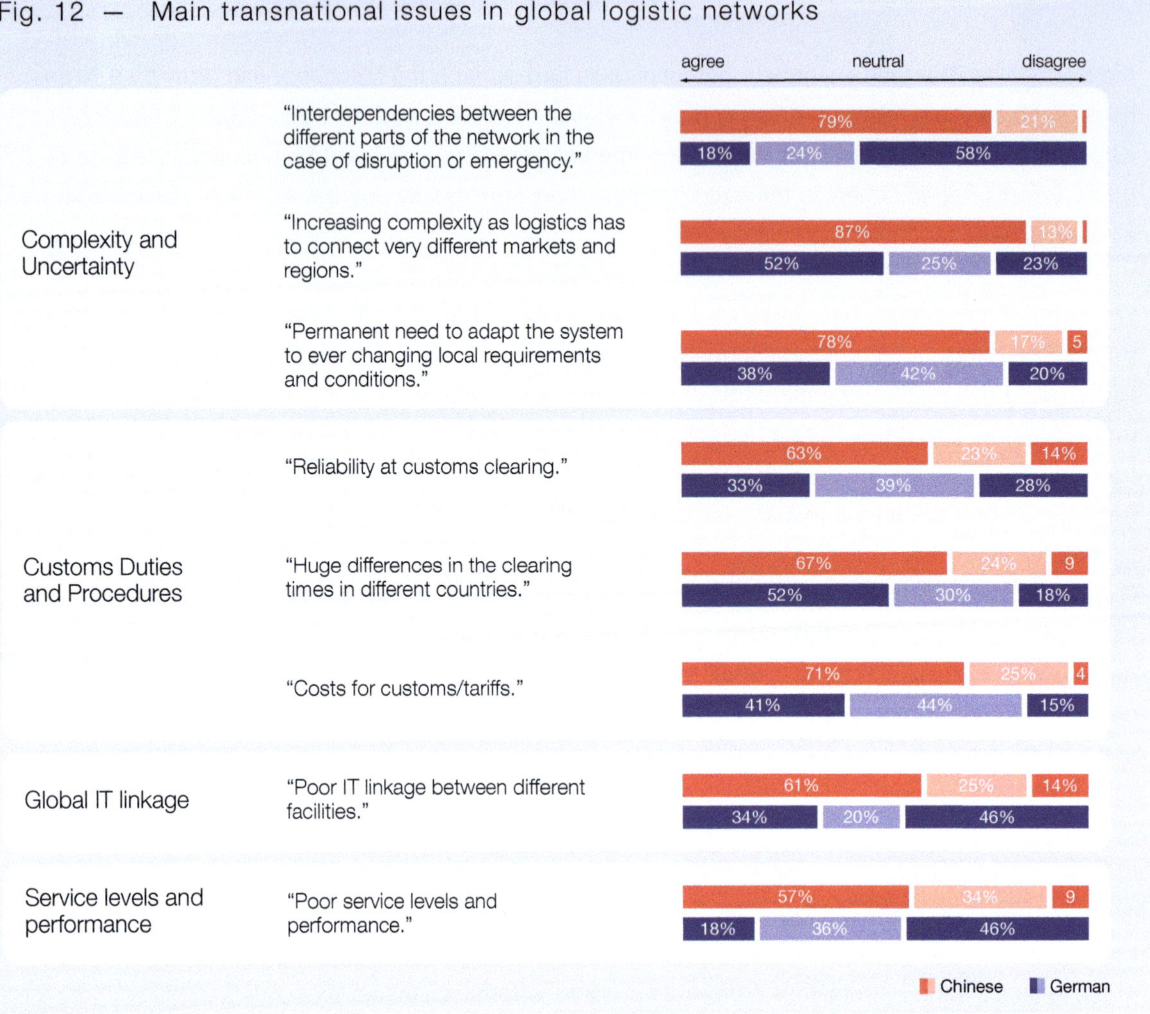

4.2 Transnational Issues in Global Logistics

Besides the country specific issues of global logistics there are a number of transnational aspects to be considered when setting up global logistics structures (see Fig. 12).

Complexity and uncertainty. Managing and executing cross-border movements of goods and information between countries worldwide make global logistics an outstandingly complex task. The complexity, comprised of long haul distances, fault-prone information exchange when linking multiple intermediaries, and diversifying external interferences, effects uncertainty in status of goods movements, higher transportation costs, risks, as well as long lead times. In international logistics networks, the risk of incorrect information exchange and misguided goods is very high.

Among others, the main reason is the high amount of persons, companies and institutions that are involved in planning, realisation and control of international goods movements.

These intermediaries can be grouped into authorities such as customs and administrative bodies, associations such as tariff unions and shipping conferences, transport companies such as airlines and shipping lines, logistics facilitators such as freight forwarders and customs brokers, and merchants such as consignees and shippers. Beside a complex documentation process, interfaces between these different intermediaries have to be bridged, which is not only challenging for physical treatment of goods but especially for directing data flow. Resulting from different software systems, information might be incorrect, not complete, or can get lost during the transfer process. Clearly, management of global logistics networks require more coordination, more communication, and the constant improvement of determined monitoring systems.

Interdependencies in case of disruptions. The complex interdependencies in today's international value chains have led to the situation that minor disruptions in one specific part of the chain may cause severe consequences elsewhere. Almost 80 % of the reviewed Chinese companies report that this as a major problem in their global networks, German companies are more confident that their risk management could mitigate such difficulties. Less than 20 % see this as an important issue.

It can be observed that in recent years the likelihood for supply chain disruptions has risen. Technological disasters, which can be defined as man-made industrial or transportation accidents such as ship loss, train crashes, and tank explosions, have increased exponentially worldwide over the last 30 years. Also natural disasters such as earthquakes, tsunamis, volcanic eruption, landslides, as well as hurricanes and tornados have become more likely to occur. As a result, they do not only cause augmented human casualties but also damage infrastructures and facilities. This can lead to temporary obstructions and losses, or at the worst, result in total destruction of global logistics networks, having profound long-term implications for firms with vendors, suppliers, or assembly facilities in the affected area.

For instance, the outbreak of the Severe Acute Respiratory Syndrome (SARS) within Southeast and East Asia in 2003 caused the decrease of local manufacture and closure of ports so that also computer assembly operations in the US had to shut down temporarily because of the lack of import components.

Customs duties and procedures. Another transnational problem is the result of trade barriers, which may be grouped into tariffs and non-tariff measures. Tariffs are various types of duties, while non-tariff measures are direct and indirect protectionism laws such as import restrictions. As customs laws, tariffs and procedures of clearance differ widely from country to country, this increases complexity and makes a separate analysis of specifications for each nation and region necessary. From a transnational perspective, this survey shows that, besides the direct cost effects of the duty itself, the time aspect of clearance procedures is almost equally important. The lack of reliability and transparency in the clearing processes in many places in the world affects the predictability in global transportation. Consequently, logisticians react with additional safety stocks and extra efforts in administration. E.g. indirect costs occur which can significantly affect the overall cost structure of international business activities. Hence, this issue may not be underestimated when planning to go global.

Global IT linkage. In global logistics, accuracy is of high importance as shipment delays and errors at a certain stage in the value chain can be tremendously amplified until final fulfilment to the customer, due the mentioned complex interdependencies. Accordingly, the intensive use of advanced information technology for planning and managing flows of goods is essential for global logistics, which includes visibility for goods in motion as well as goods at storage in warehouses and loading transfer points. In technical terms, it allows companies to focus on the status of goods and manage supply chains strategically. By looking at the supply chain from end-to-end holistically, they are able to act knowingly on information and foresee how those actions will impact a shipment's movement. Clearly, in-transit visibility brings most benefit for the shipper, but is, at the same time one of the most challenging functions to be established in global logistics. Tracking and tracing systems have to integrate a large number of involved intermediaries through a variety of means, including Web services, EDI, and legacy systems etc.

Still, the IT landscape in different countries, different companies, and in many cases even inside a firm is quite diverse. IT interfaces to link companies with their suppliers and customers in global supply chains is a key issue since years. This survey shows that one-third of the reviewed German companies and even double that of the Chinese suffer from poor IT-integration of their own global facilities.

Despite huge steps forward in terms of technology, e.g. new middle ware products and new formats for data exchange, this topic will remain a key issue in global networks in the years to come.

5 Logistics Strategies for Entering New Markets

This chapter analyses the process of foreign market entry in two steps: At first, it is looked at the role of logistics during the internationalisation process and how it is being integrated herein. Secondly, the key components of successful logistics strategies are analysed to support setting up shop in foreign markets.

Fig. 13 gives a short overview of the topics that will be discussed in more detail in the following subchapters and it will be guiding through the analysis' structure. The six different issues addressed here are considered both relevant and critical for the logistics management of a company when going global. The red and the blue line in Fig. 13 indicate how the most successful Chinese and German companies are positioned regarding each item. While the detailed analysis in the following subchapters are considering the complete sample of respondents, Fig. 13 illustrates the consolidated data from the most successful companies only, i.e. these companies have accomplished their logistics objectives in more than half of the foreign markets they operate in.

Duration of the internationalisation process (Chapter 5.1). A process model of sequential steps was developed to analyse the time aspect of internationalisation activities, starting from the initial idea to set up shop in a specific foreign market until the implementation of the new logistics system and its integration in the global network. The results show, that logistics managers at both Chinese and German companies face severe time pressure whereas Chinese companies are slightly quicker. The most successful companies take less time for setting up shop than the average.

Involvement of logistics managers (Chapter 5.2). It has been shown that the intense involvement of logistics managers in the internationalisation process from the very beginning leads to better cost and time effectiveness in the global network, and also bares the opportunity to develop additional competitive advantage. The comparison of companies from both countries indicates that Chinese companies attach even more importance to the consideration of logistics aspects and their integration of logisticians is even more intense. The influence of logistics managers at German companies is comparatively limited.

Structuring internationalisation procedures (Chapter 5.3). Chinese and German companies both consider higher levels of standardisation of the internationalisation process and its repetitive procedures as beneficial. Correspondingly, most of the successful companies have such concepts in place, including standardised checklists for market analysis, guidelines etc. to enhance the cost effectiveness and speed of the market entry and improve the characteristics of the final system. Remarkable differences can be observed regarding the application of an analytical global footprint design including methods such as Total Cost of Ownership. German companies have been gradually entering foreign markets for decades. Hence, their networks have typically grown organically over time, with less strategic planning on a global level. In contrast, Chinese companies are currently at the beginning of their global expansion. The most successful among them follow a Greenfield approach and indicate that they plan their global logistics networks right from scratch and with sophisticated and holistic planning.

Fig. 13 — Framework of internationalisation topics

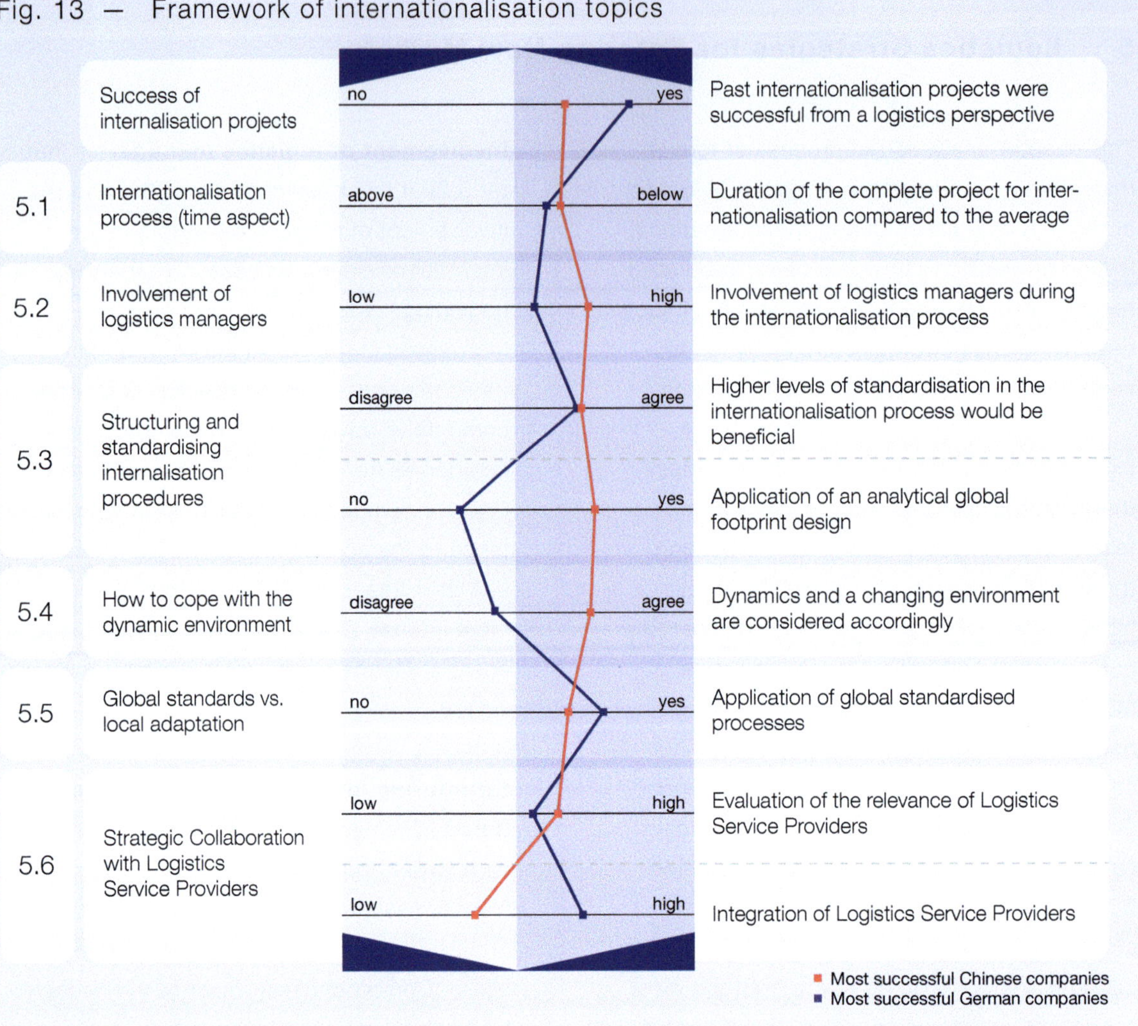

These first three sections of Chapter 5 that are dealing with characteristics of the internationalisation process are followed by the three key components of a logistics strategy for successful foreign market entry.

How to cope with the dynamic environment (Chapter 5.4). Flexibility becomes increasingly substantial for international logistics networks as they have to meet unknown future requirements and are operated in an ever changing business environment. Hence, it is analysed to what extent flexibility aspects are incorporated in logistics strategies for going global. Among the most successful companies of the sample, the Chinese consider flexibility objectives to a higher degree than German companies, who focus more on quantitative aspects like costs and speed in their systems.

Global process standardisation (Chapter 5.5). Decisions in the process design of international logistics systems are to be made between the two extremes of global standardisation, in order to reduce complexity, and local adaptation, to meet local peculiarities in logistics conditions and customer expectations. The majority of respondents appreciate the benefits of globally standardised procedures, though its application remains limited. The most successful companies indicate higher levels of global standardisation than the average – this is especially the case for German companies, where the discrepancy between the most successful companies and others is especially severe.

Strategic collaboration with logistics service providers (Chapter 5.6). When companies enter foreign markets they show higher outsourcing levels and the range of activities that is outsourced to LSPs is increasing. Chinese and German respondents consider the cooperation with LSPs as similarly beneficial; it is regarded as a strategic topic of global logistics management. Despite the fact that the most successful Chinese companies value the collaboration with LSPs as even more important than German firms, they ultimately outsource to a lesser extent. This is most likely due to the fact that outsourcing in China has not yet reached the same diffusion as in western economies. Hence it can be expected that these companies will outsource more in the future, when the general outsourcing culture of these upcoming global players will have altered.

All these aspects will be examined in more detail in the following sections and Fig. 13 will be of help in navigating through Chapter 5.

Insight — Eastern Europe

After the end of the Cold War and the fall of the Iron Curtain in 1989, Eastern Europe was the scene for numerous reforms as well as the transition from planned economies to open markets. In this survey, Eastern Europe includes Bulgaria, Croatia, Czech Republic, Estonia, Hungary, Latvia, Lithuania, Moldova, Poland, Romania, Slovakia, Slovenia, and the Ukraine.

As of 2006, Eastern Europe made up 2.4 % of the world's population and 1.5 % of the world's share of GDP compared to 6.0 % (population) and 24.5 % (GDP) in Western Europe. However, Eastern Europe has received much investment interest in recent years, partly because, as a region, it has been labeled a Rapidly Developing Economy. Economies in Eastern Europe are characterized by consistent GDP growth, growth of industrial output and decreasing unemployment rates. Some of the countries have made huge efforts to catch up with technology and modern lifestyle. In Estonia, for instance, ministers hold paperless meetings on laptops and a high percentage of residents files their tax returns by e-mail. However, heterogeneity can be observed thoughout Eastern Europe. The downside of the emerging markets begins just a few kilometres beyond the sparkling new skyscrapers: decaying villages with gray and peeling facades, many without paved streets. In small villages in Poland, for instance, nearly half of the population is out of work. Other members of the former Eastern bloc also have their poorhouses, where time seems to have stood still. The gulf between rich and poor has widened dramatically in the past few years.

The 2004 EU enlargement included East European countries, which were the Czech Republic, Estonia, Hungary, Latvia, Lithuania, Poland, Slovakia, and Slovenia amongst others. Again in 2004, Croatia became the status of a candidate country; Moldova and the Ukraine have future enlargement possibilities. In 2007, Bulgaria and Romania joined the EU. In comparison, Eastern and Western Europe are extremely diverse, not only in terms of culture and language, but also in terms of economic and industrial strengths and weaknesses. Nowadays many West European companies, for example, apply the concept of near-shoring to East European countries due to comparative cost advantages. One of such an advantage can be realised with the ongoing EU accession of Eastern European countries where low-wage workers offer their manpower to western economies. Overall, Eastern Europe's history has seen many changes, making it home to many different cultures and languages, and thus a strongly heterogeneous region. As the EU further enlarges, historical borders that separate nations will be considered as obsolete, as the entire European market will be harmonized, including transport and logistics infrastructure.

Today, railway systems in Eastern Europe, traditionally expanded over decades, represent the most important transportation mode and therefore have a high diffusion in terms of intra-regional logistics services. When looking ahead, however, transportation will shift from railway to road on a long-term effect such as it is being observed in Western Europe, for example. For the supply of Eastern European companies, short sea shipping is comparatively important. Although, harbours in Eastern Europe are not ranked among the most important in the world, short sea shipping is still usual for loading and delivery, and harbours at the Baltic Sea benefit from European sea transportation. However, compared to Western Europe, the Eastern part of Europe still shows poorly developed construction of roads, railways, and airports.

Before the necessary reforms, the Eastern European logistics sector was affected by state-run logistics companies that featured rigid and given market structures. According to the economical reversal, the logistics market now is reseeded by numerous small logistics service providers comparable to the situation in China after market liberalization. Actually, none of the European logistics service providers is capable of supplying whole Europe with self-run operations, not even globally acting providers from Western Europe. Herein, there is more room for expecting the prospective Pan-European network. The most important reason for a non area-wide and continuously developed infrastructure is the heterogeneity amongst the member states in Eastern Europe. However, in recent years the so-called "Pan-European Transport Corridors" between eastern and western member states were established with the objective to facilitate transportation amongst different countries and regions. Herein, the major difficulty was and will be financing the mandatory investments. Even though the European Union agreed on financial support, the main part of the funding remains with the individual member state.

When it comes to international logistics services, the influence of Western Europe is still remarkable. Shipped goods and commodities from Asia and North America, for example, are directed to West European harbours like Rotterdam, Hamburg, Antwerp, and Bremen, also known as the North-West-Range, before they are transported to their destinations in Eastern Europe via short sea shipping, by rail, lorry, or aircraft. Therefore, equally important for supplying the eastern part of Europe are major cargo airports in Western Europe such as Frankfurt, London, Paris, or Amsterdam. Hence, international flow of commodities and haulage into the eastern part of Europe is mostly transferred via Western Europe.

Due to the political and economical changes resulting from the EU enlargement, new highly developed hubs are continuously emerging and many distribution centers are being relocated to the eastern part of Europe. Black Sea harbours and the Trans-Siberian Railway gain significance for international transportation, due to time and cost savings, especially concerning the linkage from and to Asia. Locations such as Bratislava (Slovakia), Constanza (Romania), Odessa (Ukraine), and Katowice in Poland are important logistics locations with huge potential. With a population of more than 38 million, for example, Poland as the largest new EU member is both an affordable production site and a market full of demanding consumers.

As a matter of course, heterogeneity also affects logistics markets. Quality and performance of logistics services are unevenly spread among the East European member states. On the one hand, advanced logistics markets with international and globally acting logistics service providers exist, on the other hand, poorly developed logistics structures are not uncommon. Poland, Slovakia, Hungary, and Czech Republic are numbered among the advanced economies regarding logistics structures. Until the end of 1990s, for instance, Slovakia was considered the problem child of Eastern Europe, holding up negotiations for EU membership. Today, foreign financiers value the country as an investment jewel. A notable foreign investor is Kia, for example, who launched its production in late 2006, giving Slovakia a huge boost to her automotive industry. The plant is projected to bring 10,000 new jobs. Kia was not the first auto manufacturer who discovered the remote Eastern European country. VW produces its SUV, the Touareg, in Bratislava. Hence, Slovakia is among the EU's most attractive countries for foreign investors, due to its low labour costs and moderate tax rates. The other side of the coin are countries such as Hungary and Romania that are poorly developed in terms of logistics due to their geographical position and industrial structure. Though, due to low labour costs, an increasing volume of inward processing can be observed here. In between, the Baltic countries extend their potential as important hubs between Europe and Russia.

5.1 Duration of the Internationalisation Process

In the rapidly changing environment of today's globalising markets, time is continuously gaining importance and, as a matter of fact, timing is considered to be an essential part of any business strategy in the context of foreign market entry. Thereby the time aspect has several dimensions: one is the specific point in time, when the market will be penetrated, and second is the duration of the complete process of foreign market entry. To cater ambitious timing strategies, to keep the costs of a market entry low and to utilise assets as quickly as possible, internationalisation projects suffer from tremendous time pressure. Chinese companies state that they set up shop in foreign country within 14 months, German companies need on average almost 18 months.

Structure of the internationalisation process. For this survey, the internationalisation process is defined to start with the initial strategic idea of entering a specific foreign market, and is finished with the completed implementation of the new local logistics structure and its integration into the global network of the respective company. In general, such a project comprises several steps on all decision making levels of a company. It can be understood as a top-down process, with top-level decisions in the beginning, handed over to middle and lower management levels, and operational tasks in its final stages.

Fig. 14 illustrates a simplified internationalisation process that was developed for this research. It consists of four sequential steps that represent the main phases of a typical foreign market entry from a logistician's point of view. These are: (1) evaluation of the new market, (2) strategic planning, (3) logistics planning, and (4) implementation of the logistics system.

Of course, the complete process in real business practice is much more complex and involves many different functions of the company in one way or the other; though the simplification of the process permits an analysis of the time aspect of such projects with regard to the internationalisation of logistics networks.

Step 1: Evaluation of the new market.
Triggered by the initial strategic idea to set up shop in a specific market or region, the internationalisation process begins with comprehensive market analysis. In addition, the top-management level evaluates the overall conditions in the target market: sales potential (market size and potential market share), input factors (e.g. local supplier base, cost for labour, resources and land), infrastructure for domestic and international transportation, political and legal aspects etc. The set of objectives for the market entry is defined (comp. Chapter 2) and different designs of thinkable business cases are examined. If the analysis shows that the market entry is feasible and it is expected to meet the respective objectives, this first phase is finished with the final decision to set up shop.

For some companies the initial idea results from a continuous monitoring of internal KPIs and country specific indicators, which gives a signal whenever a market entry seems to be favourable.

With about one-third of the total project's duration for both German and Chinese companies, this phase accounts for the largest share, compared to all other phases. As so many aspects need to be considered and evaluated, it is clear that the expertise and know-how of almost all corporate functions need to be included to create a holistic and realistic picture of the market.

Fig. 14 — Duration of the internationalisation process

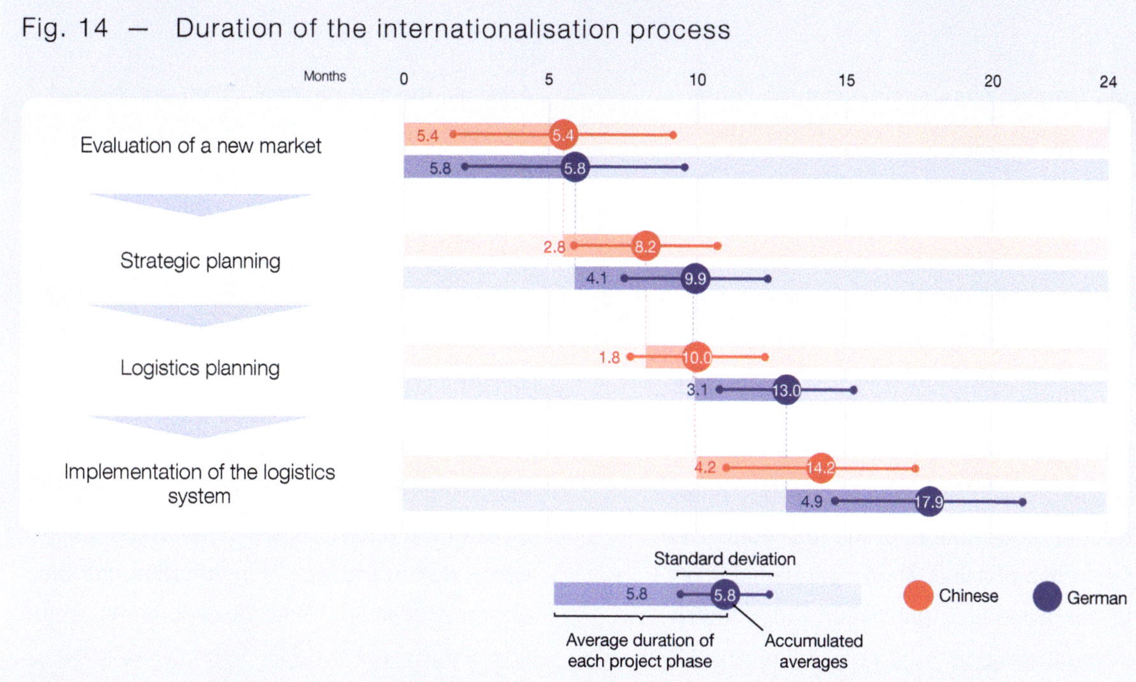

On average, German and Chinese companies require between 5 to 6 months; Chinese companies seem to be slightly quicker.

Step 2: Strategic planning. The final decision to set up shop in the market is followed by the phase of strategic planning. Still on the top management level, in this phase the foreign market entry is outlined as a project, with a time schedule, defined milestones, budgets, and concrete objectives for the different corporate functions. Cross-functional aspects of the planning process will be clarified and concrete measures are prepared to ensure that the functional planning afterwards will not create unexpected interferences.

The result of this phase is a framework, which will guide the functional divisions of the company to develop their plans of action according to the strategic objectives of the foreign market entry.

This phase is by far shorter than the evaluation phase. Chinese respondents calculate less then three months for this phase, Germans with slightly more than four months.

Step 3: Logistics planning. When the strategic planning is finished, the different corporate functions involved, including the logistics department, start their specific planning. In close interaction with all other relevant functions, inbound, outbound, and in-house logistics processes are defined and implementation plans are developed. Relevant topics in logistics planning include the definition of service levels, intended lead times, inventory policy, network structure, capacity calculation, allocation of facilities (e.g. warehouses, cross docks), IT integration, decisions about logistics outsourcing, and preparation of tenders.

The phases of strategic and logistics planning are relatively short in comparison to the other two. At Chinese companies, logistics planning accounts for less than two months, at German companies it takes slightly more than 3 months. This reflects the severe time pressure under which these complex logistics structures are being planned.

Step 4: Implementation of the logistics system. After 10 months Chinese companies start implementing the new logistics system along with its integration into the global network. German companies begin the implementation on average three months later.

Being extremely complex and crucial for the internationalisation's success at the same time, the implementation phase is comparatively time consuming. With 4.2 months at Chinese and 4.9 months at German companies it is the second longest lasting phase of the foreign market entry. Here it shows, if the plans and decisions made in the early phases meet the requirements of the real world. The work in this phase is at an operational level and it requires an organisation with comprehensive problem solving skills, as unexpected occurrences and developments cause a lot of fire fighting. The global integration of processes, people, and technology is extremely challenging and it is important that the people who are working on this are well trained, not only on logistics, but also in leadership and cross cultural management, and that they are empowered to take decisions at the level where they are required.

Characteristics of the process. On average, Chinese companies indicate that the complete internationalisation process lasts for approximately 14 months. German companies exceed that with a mean duration of almost 18 months. As stated in Fig. 13, the most successful companies are below average. Hence, they cope with the challenges of time pressure more efficiently than the rest of the sample.

After all, the internationalisation process addresses aspects of cross-functional as well as the cross-company integration from the top management to operational levels, from the initial idea of going global to the point of implementation. In reality, the process is by far more complex than the sequential model that was applied here. Typically, several steps are carried out concurrently and the interdependencies are hard to overlook. Especially at the beginning of the internationalisation process, most of the decisions are made under uncertainty, due to the lack of information and the unpredictability of future developments. In the final stages, planning is of less importance while pragmatism and incrementalism are required to get the market entry done.

Insight — Useful Indicators for Country Assessment

Logistics services differ significantly in their quality and performance across countries and regions. In Kazakhstan, for instance, it takes 93 days to export a 20-foot full container load (FCL) container of cotton apparel, and in Mali 67 days, while in Sweden it only takes 6 days. These variations in time and costs respectively across countries and regions result from differences in quality and costs of infrastructure services as well as differences in the economy, culture, policy, etc. Overall, all these aspects have a significant effect on trade competitiveness.

After the initial strategic idea to set up shop in a foreign market or region, the evaluation of a new market builds up the first step in the internationalisation process (comp. Chapter 5.1). As a result of globalisation, it has become increasingly challenging to identify internationalisation strategies, to choose which countries to do business with, and where to locate an offshore entity. Among basic macroeconomic figures like GDP, GDP growth, per capita income growth rates, purchasing power parity, exchange rates etc., many more indicators have to be taken into consideration in order to provide a first assessment and orientation whether or not to enter a foreign market.

Macroeconomic figures alone cannot suffice when entering new markets, especially if the country's background, history, culture, infrastructure, habits, and policies differ greatly from the home economy. To analyse a specific country or region from a logistics perspective, figures such as transportation costs, timeliness of shipments, transport and IT infrastructure, customs and border procedures, and the overall logistics competence of foreign markets have to be taken into account.

A variety of institutions continuously aggregate facts and figures to meaningful country indices on a mutual basis. In order to facilitate a quick start for logistics managers in internationalisation projects, a selection of helpful indices for country assessment is stated here:

Global Competitiveness Index (World Economic Forum – www.weforum.org)

The Global Competitiveness Index by the World Economic Forum provides a holistic overview of factors that are critical to driving productivity and competitiveness, and groups them into nine pillars: institution, infrastructure, macroeconomics, health and primary education, higher education and training, market efficiency, technological readiness and business sophistication. The higher the respective country's Global Competitiveness Score, the more competitive the economy.

Economic Freedom of the World (The Fraser Institute – www.fraserinstitute.ca)

The index published in Economic Freedom of the World by the Fraser Institute measures the degree to which the policies and institutions of countries are supportive of economic freedom. The cornerstones of freedom are personal choice, voluntary exchange, freedom to compete, and security of privately owned property. Thirty-eight data points are used to construct a summary index and to measure the degree of economic freedom in five areas: size of government, legal structure and security of property rights, access to sound money, freedom to trade internationally, and regulation of credit, labour, and business.

Hiring and Firing Indices (World Bank – www.doingbusiness.org)

The Hiring and Firing Indices by the World Bank's Doing Business measure the flexibility of labour regulations. It examines the difficulty of hiring a new worker, rigidity of rules on expanding or contracting working hours, the non-salary costs of hiring a worker, and the difficulties and costs involved in dismissing a redundant worker. Higher values of the respective indices indicate more rigid regulations. Further indicators by the World Bank's Doing Business include topics concerning tax payment, contract enforcement, license handling, and investor protection.

Corruption Perception Index (Transparency International – www.transparency.org)

Transparency International's Corruption Perception Index is a survey-based index that ranks the degree of corruption as seen by business people and country analysts. It ranges between 10 (highly clean) and 0 (highly corrupt).

Logistics Perception Index (Global Facilitation Partnership for Transportation and Trade – www.gfptt.org)

The evaluation of the Logistics Perception Index is still underway. It is based on a survey on the logistics environment of countries gathered from managerial level personnel of international freight forwarding firms worldwide. The survey is supposed to help generate an informed set of logistics perception indices that measure the key dimensions of logistics for developing and industrialized countries. Aspects of measurement are economic impact of trade facilitation, trade logistics and developing countries, trade logistics and practical measures and trade logistics strategies.

Global Logistics Indicators (World Bank – www.worldbank.org)

The World Bank has determined an array of Global Logistics Indicators to measure the efficiency of international trade procedures. Here, characteristics of the global environment, respectively country-specific conditions such as differences in the quality of infrastructure services as well as differences in policies, procedures, and institutions are direct subject to the indicators. Although primarily designed to help governments in deciding actions towards improvement of logistics performance, the World Bank's Global Logistics Indicators can also give a guideline for logistics professionals. They provide a framework of classification and criteria that can help managers attain an overview of their respective fields of interest in country assessment from a logistics viewpoint. The global logistics indicators can be classified in different dimensions of measurement, which are primarily cost, time, complexity, and efficiency. This classification provides a starting point for the manager in a more advanced stage of target country/region analysis. A possible set of global logistics indicators is outlined in Fig. 15.

Fig. 15 — Global logistics indicators structure

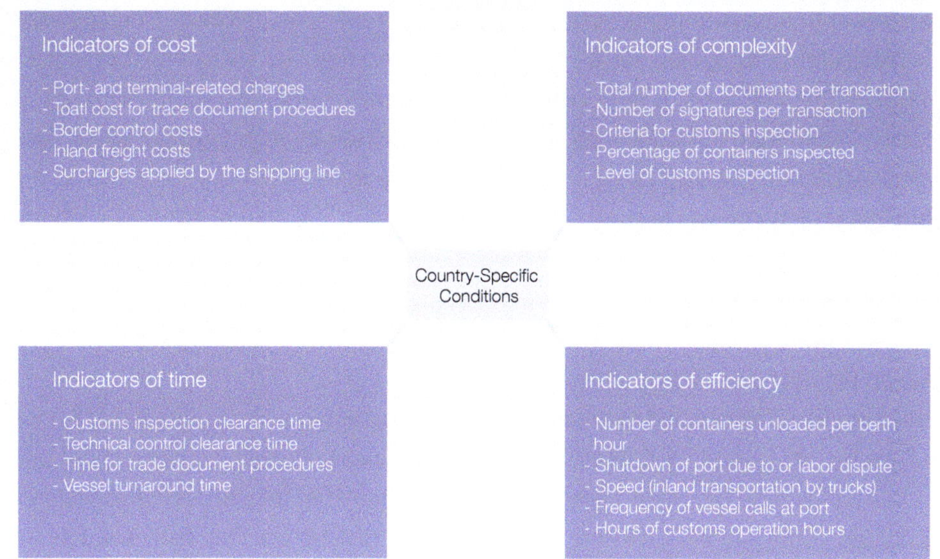

As a result, all of the different indices can be compared against macroeconomic indicators and allows benchmarking across countries and regions from a logistics perspective considering various indicators of development levels. Different indices should also be independently taken into account when operating in different countries. Furthermore, other global logistics indicators also have to be measured according to each specific country when evaluating the quality of business environment and logistics services. Due to different projects of trade facilitation being initiated by international associations as well as national and local authorities, metrics of global logistics indicators are changing in each country and have to be updated regularly. These indicators help managers assess logistics performance in the country of interest, providing sensitisation for logistics performance for each respective target country.

5.2 Consideration of Logistics within the Internationalisation Process

In order to benefit from the advantages that motivate companies for going global, logistics costs and performance levels have to be taken into consideration from the initial idea for internationalisation until the final implementation and integration of new facilities. This leads to the question to what extent logistics and logisticians are being integrated institutionally or methodologically in the internationalisation process and whether there are differences between Chinese and German companies. Thereby two factors have been considered: the time aspect, i.e. when logistics and logisticians come into play, and the degree of their overall involvement during the internationalisation process.

Importance of an early integration of logistics. One significant issue to ensure the success of new market entries is to consider logistics and integrate logisticians at a very early stage of the process (comp. Chapter 5.1), i.e. during the evaluation phase and the strategic planning. Unexpected high logistics and transportation costs frequently outweigh the advantages that motivate for going global from a manufacturing, procurement, marketing or sales perspective and hence endanger the entire foreign investment. Already at the stage of defining the basic market entry strategies and designing the respective business case for the new region, it is mandatory to take the opportunities and expected shortcomings of the future logistics system into account.

A look at the most successful projects shows that comprehensive analysis of the logistics conditions and feasibility studies for several alternatives of logistics systems do not only lead to enhanced cost structures of the final logistics solution. Furthermore it is possible to develop significant competitive advantage by achieving superior customer service.

More than two-thirds of all respondents stated that late involvement of logistics during the internationalisation process will lead to unnecessary high logistics costs (see Fig. 16). More than half of them see this as a potential threat to the market entry as a whole. The results show that Chinese managers take this issue more serious than their German counterparts.

Despite these clear statements from the logisticians interviewed here, about one-third still feel that they are not involved in the decision making process early enough — which shows that not all companies do utilize logistics know-how to establish a high level of transparency in the cost structure of global networks as well as to create competitive advantage.

Degree of logistics involvement in the decision making process for new market entries. The extent to which logistics managers are involved in the decision making process differs widely, and there are huge differences between German and Chinese companies. While 60 % of the Chinese respondents state that logistics is always being considered when it comes to the decision to enter a new market or not, less than 30 % of their German peers experience this level of involvement.

Concerning three specific topics of internationalisation, it was investigated as to whether logisticians are integrated in the decision making or not: selection of locations for (1) distribution centres/logistics facilities, (2) manufacturing facilities or outlets and, (3) suppliers. While most of the respondents are, by nature, involved in locating logistics facilities, German companies still show little consideration of logistics in the other two issues. For Chinese companies more than two thirds of the respective logistics managers get

Fig. 16 — Involvement of logistics managers in internationalisation projects

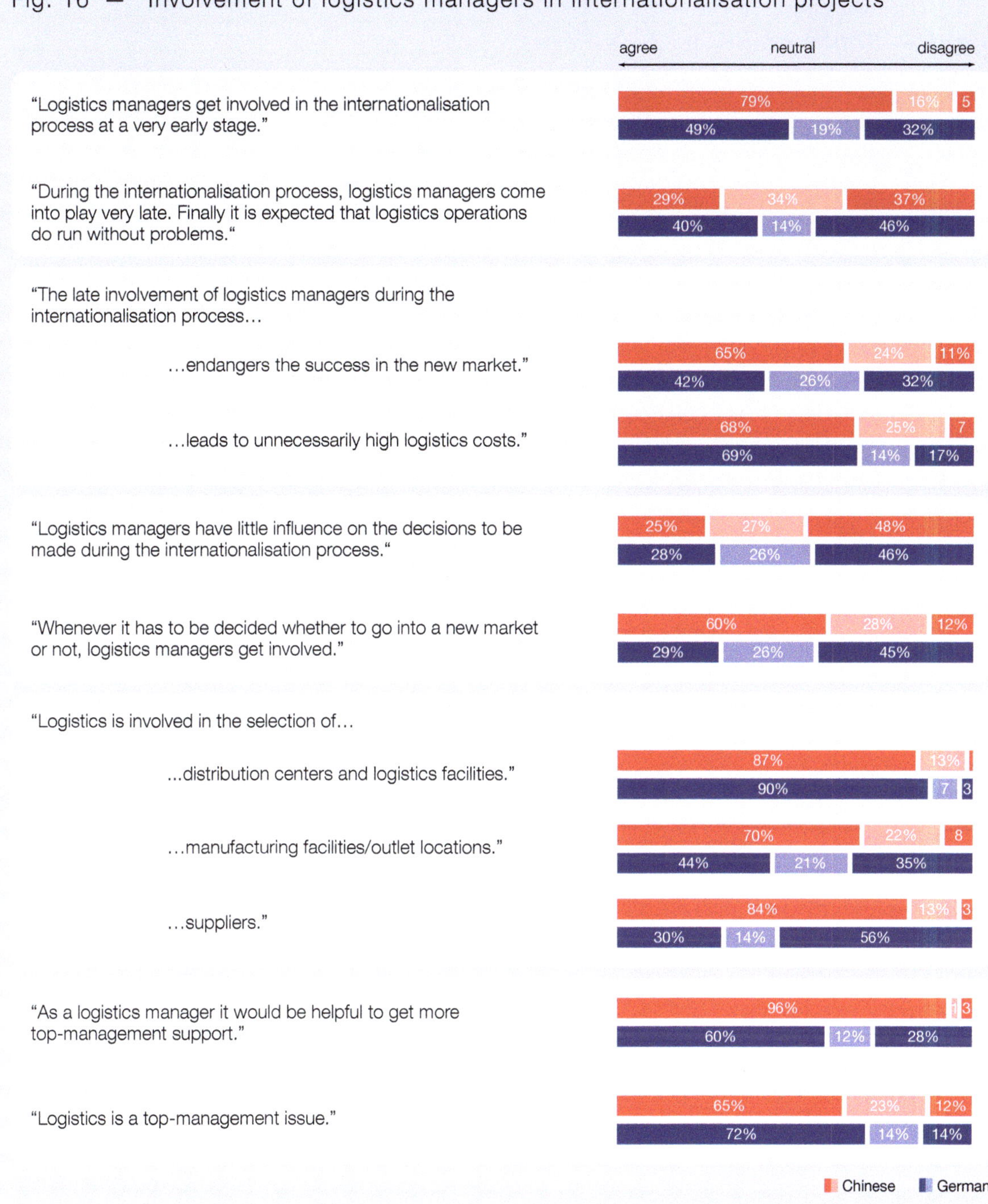

involved in locating manufacturing sites and outlets as well as in supplier selection. However, in more than 25 % of the total reviewed companies, logistics managers have little influence in the final decision making. As mentioned before, the opportunity to benefit from methods like Total Cost analysis to establish the transparency of costs among several partners in a value chain by leveraging logistics knowledge is still widely unused.

This shows that a lot of companies today do not yet use the full potential of logistics as a holistic, integrated tool for customer and network management. Thereby, the strategic potential of logistics for the international business success remains in many cases unleveraged. That is why almost 70 % of the total respondents in this survey call for more top management recognition and support to avoid miscalculations, prevent follow up costs, and suboptimal international logistics networks.

Insight — South America

South America includes the 12 countries Argentina, Bolivia, Brazil, Chile, Colombia, Ecuador, Guyana, Paraguay, Peru, Suriname, Uruguay, and Venezuela with a population of 373 million people. Because of lower levels of competition and higher unmet demand, the vast region is now regarded as perhaps the second most lucrative international market behind Asia. South America is responsible for 4 % of the world GDP whereas the region forms 5.8 % of the world population.

Starting in the 1980s, democratisation has found its way into a multitude of South American parliaments. The increased democratisation of governments there is promising more stable long-term investment opportunities. At the same time, except for certain cases, South American countries are moving away from state-owned companies monopolizing the industrial sector, and embracing foreign investment and joint ventures as an effective way to boost development. Nonetheless there is still a certain number of socialist leaders governing the continent with a tendency to refuse free market capitalism.

An example of the economical instability is the Argentinean Crisis at the end of the last century. Argentina experienced a major economic collapse due to high interests on debts and an overvalued Peso pegged to the American Dollar. The result was reflected in the high unemployment rate of 23 % and a devalued currency of 30 %. The rate of the population under the poverty line reached 57 %. As international markets are more interconnected than ever before, financial crisis like in Argentina are quickly felt in neighbouring countries like Brazil and Venezuela. The South American continent still has not fully recovered from the setback.

Many problems remain to be solved in South America. The South American market is an export and import market; only 11 % of the Latin American trade is internal trade compared to Intra-European trade (EU 25), which accounted for two-thirds of Europe's exports in 2005. This minimal internal trade is mainly limited to the larger cities. The purchase power is highly concentrated in urban areas. In fact, 25 % of the South American population is residing in the ten largest cities. Some 80 % of the population in Latin America consists of low income sectors. The unemployment rate in South America with more than 10 % is respectively high and the inflation rates are twice as much as in the USA. The economic gap between rich and poor is leading to enormous descrepancies.

The complexities in tariff rules, customs procedures and practices in most countries in South America are relatively high and therefore, for example, implicate difficulties in providing logistics services. Taxes and licensing provisions typically increase overall costs for private investors, and therefore impede private investments. Tax burden is very high, with the state commanding 90 % or more of total profits in Venezuela, for example. Therefore, the design and structure of tax systems in each country affect the extent to which financial risks are borne by private investors. Interest rate fluctuations, occurring currency depreciations and political instabilities should also be taken into consideration.

Despite all the economical and political risks, the 12 South American countries have agreed on the foundation of an intergovernmental union in 2004, so-called "Union of South American Nations", similar to the role model of the supranational European Union. Furthermore, two existing free trade organisations have been founded long before, the Andean Community of Nations in 1969 and Mercosur in 1991, whose purposes are to promote free trade and the facilitated movement of goods, people, and currency. Since the founding of these organisations, trade barriers were dismantled significantly and therefore trade has increased significantly. In addition, a harmonisation of laws is intended in order to have positive implications on intra-regional and international trade. These aspects are promising for a more open market access and therefore a greater trading volume.

From a logistics perspective, the major challenges are related to infrastructure. The lack of a modern infrastructure system and reliable communication networks leads to a low level of performance accompanied with high logistics costs. These remain much higher than the standard, with a share of 12-20 % in total product costs compared to only 6-10 % in more developed regions. Most of the international movement of goods are transported by aircraft, while other international haulage is processed by sea freight. Regarding sea and air freight, however, none of South American harbours or airports is listed among the global top 30 in terms of turnover. Globally acting logistics service providers, for example, mostly serve their South American customers from North American hubs, e.g. Miami (USA). Railways are rarely being used for transportation because of the geographical and topographical difficulties they bare. The geographical challenging Amazon, Andes, and rain forests are insurmountable by rail. Besides, apart from not being modern, railways are not uniform from region to region and the terminals are currently not equipped to unload or reload goods.

A recent Georgia Tech study showed that 67 % and 72 % of companies surveyed in Latin America used third-party logistics providers in 2004 and 2005, respectively. Of these, 77 % rated the relationships with 3PLs as very successful. Logistics service providers regard South America as a challenging market as they have to deal with political and economic instability, port and transportation infrastructure challenges, and regulatory and tax issues. The major share of the growing logistics market in South America is driven by FDI as well as the 10-20 % annual growth of demand for international logistics. Although the logistics market in South America is growing, there are many problems that have to be solved. One of these problems, for example, is related to the logistics processes of domestic industry enterprises. They lack to some extent basic logistical concepts in the form of replenishment and shipment tracking.

The current trend to bring South America's economy up is an increase on the FDI side. This is forcing domestic firms to compete and cut costs, while bringing South America into the global network of suppliers and customers.

5.3 Structuring and Standardising Internationalisation Procedures

Internationalisation in the sense of setting up shop in a foreign market is not a unique activity for a firm. Internationalisation projects are typically carried out frequently, with different target regions. Hence, it is analysed, to what extent these projects and their repetitive procedures are being unified by logisticians, or if they treat every market entry separately. It shows that some logisticians take each individual market entry as a completely new project, with little standardisation in its procedures.

Afterwards, it is analysed whether companies apply a logistics strategy for setting up shop in foreign markets and which objectives such a strategy follows.

Standardisation of the internationalisation process. Looking at the internationalisation process model suggested in Chapter 5.1, it becomes clear that certain procedures will occur in any such project, regardless of the respective target country. Market evaluations, feasibility studies, analysis of the country specific infrastructure, methods for the location of facilities, heuristics and routines for network design and many more repetitive activities, could be standardised, to enhance the quality and reliability of these procedures.

Standardisation in this context is seen to be highly beneficial. More than seven out of ten respondents conclude that higher levels of standardisation would enhance the quality, the speed, and the cost effectiveness of internationalisation processes (see Fig. 17).

Despite these clear statements, not all companies have effectively implemented such concepts. 45 % of the Chinese and 34 % of the German managers indicate that their companies' internationalisation process is not yet sufficiently standardised. Though, the majority

of the respondents call for more standardisation in the future (see Fig. 17).

Standardisation is valuable, but has its limits. The complex heterogeneity between the different countries and regions makes it necessary to treat every foreign market individually. The standardisation approach has to be understood as a generic tool that provides a comprehensive framework for the market entry, and one that is not too detailed and leaves sufficient room for individual adaptation. The evaluation of the most successful companies of the sample shows that they do benefit from higher levels of unification in their internationalisation procedures (see Fig. 13). Still, 17 % of Chinese and 37 % of the German logisticians neglect the general applicability of standardisation because of the severe distinctions between countries.

Companies with standardised procedures typically have unified guidelines and handbooks in use, provide methods and checklists for market evaluation and establish a special team of experts to support any internationalisation activity with know-how and expertise (see Fig. 18). Such teams include, in more than half of the cases, logistics managers. No matter whether standardisation is in place or not, any company should make sure that it learns from former market entries and leverages the respective knowledge for future projects.

Just like any standardisation has to leave room for adaptation when needed, it is also clear, that not all decisions during the internationalisation process can be grounded on precise analysis. In contrast, 44 % of the Chinese and 20 % of German respondents state that in many cases, the individual "gut-feeling" of managers is of significant importance. Strategic and analytic planning has its limita-

Fig. 17 — Status quo and benefits of standardised internationalisation procedures

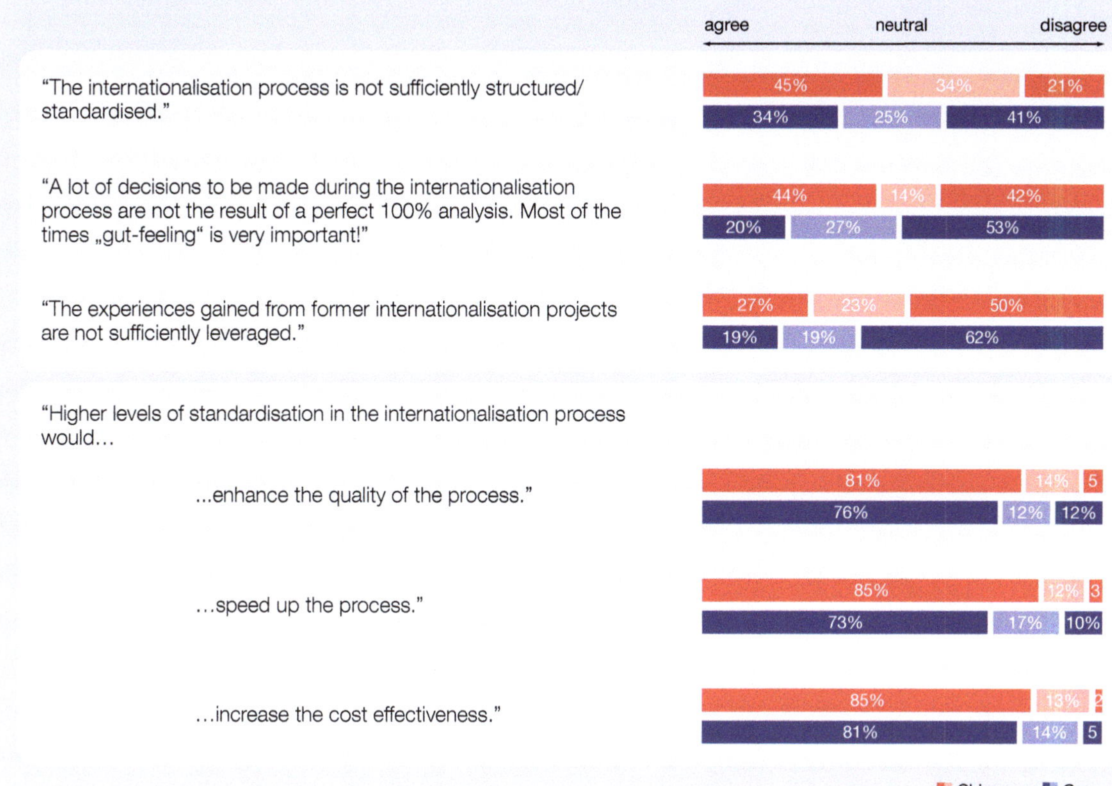

tions and especially in the final stages of the internationalisation process, pragmatism and common sense for improvisation is required to get the job done efficiently.

In general, the data shows that the share of German companies who apply higher levels of standardisation is slightly higher compared to Chinese companies, who on the other hand have higher expectations on future standardisation. This could be explained by more international experience at German companies, and having already accomplished comparatively more market entries in the past.

Application of a logistics strategy for internationalisation. With the development of logistics towards an integrated corporate function (comp. Chapter 3.1) and its increased importance for corporate success, the concept of a logistics strategy has emerged, as a subsystem of the corporate strategy. Along with the standardisation of internationalisation procedures, this survey figured out whether companies apply a strate-

gic logistics approach for going global.

About half of the Chinese and two-thirds of the German companies have a logistics strategy for internationalisation in place, consisting of strategic objectives, long-term plans, and policies. Most of the logistics managers get instructions or performance objectives from higher hierarchy levels, e.g. from the top management during the strategic planning phase (comp. Chapter 5.1). In many cases logisticians are expected to meet the same performance and cost levels like in their home market (see Fig. 19)

The approach of those companies that do not apply a logistics strategy and do not engage in strategic planning when entering foreign markets, is best described with the concept of Incrementalism or Muddling Through. In order to operate in the global environment with its complexity and rapid changes which can hardly be covered with strategic plans, these companies enter foreign markets in a sequence of incremental steps. This enables

Tab. 3 — Logistics objectives of foreign market entry

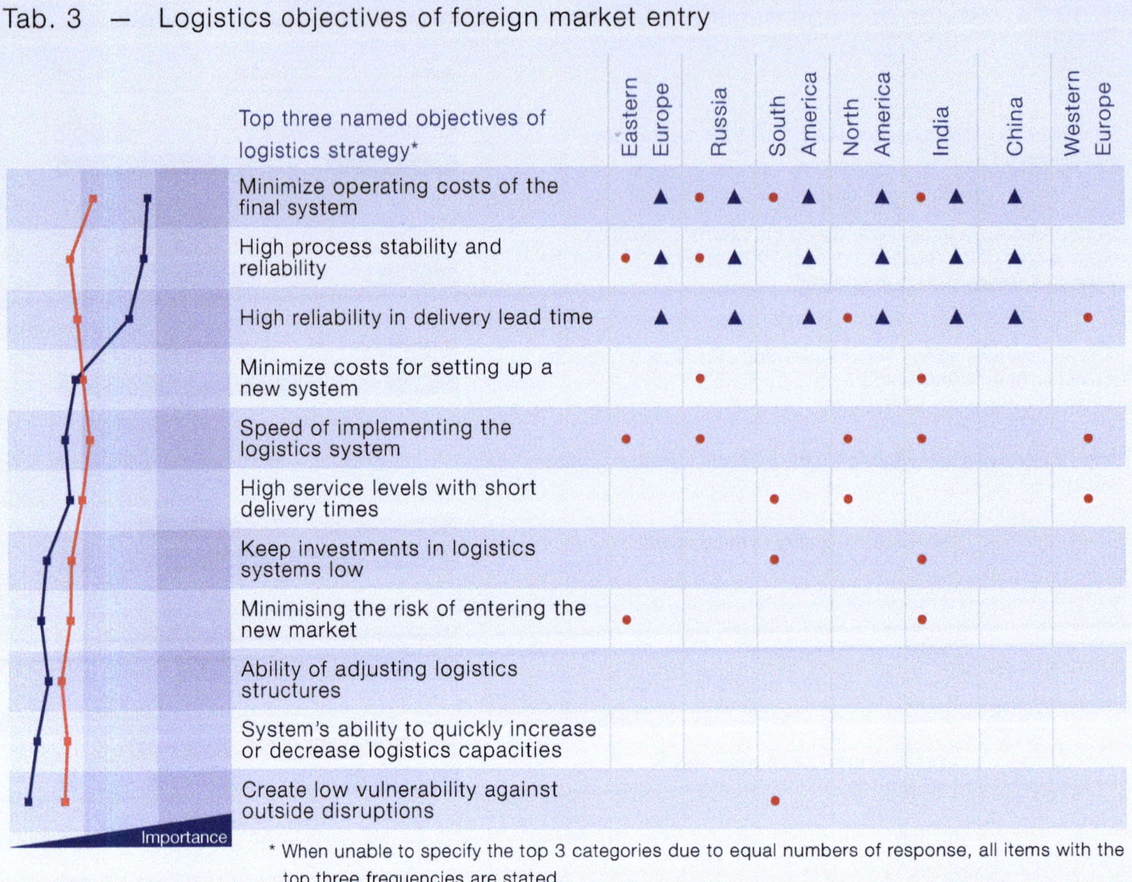

Top three named objectives of logistics strategy*	Eastern Europe	Russia	South America	North America	India	China	Western Europe
Minimize operating costs of the final system	▲	● ▲	● ▲	▲	● ▲	▲	
High process stability and reliability	● ▲	● ▲	▲	▲	▲	▲	
High reliability in delivery lead time	▲	▲	▲	● ▲	▲	▲	●
Minimize costs for setting up a new system		●			●		
Speed of implementing the logistics system	●	●		●	●		●
High service levels with short delivery times			●	●			●
Keep investments in logistics systems low			●		●		
Minimising the risk of entering the new market	●				●		
Ability of adjusting logistics structures							
System's ability to quickly increase or decrease logistics capacities							
Create low vulnerability against outside disruptions			●				

* When unable to specify the top 3 categories due to equal numbers of response, all items with the top three frequencies are stated.

● Chinese   ▲ German

Fig. 18 — Standardisation of internationalisation procedures

Fig. 19 — Logistics strategy for foreign market entry

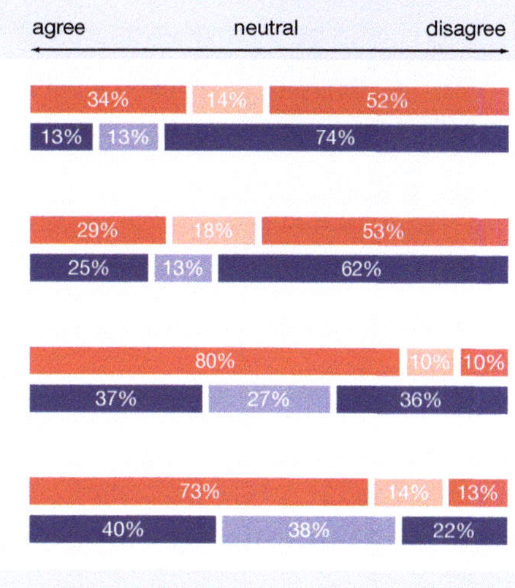

the company to quickly adapt to unforeseen developments, as problems are immediately handled at the time that they occur. As internationalisation procedures suffer from high uncertainty und unpredictable developments, incrementalism seems to be the choice for quite a few companies.

The comparison of German and Chinese companies indicates that the application Muddling Through for the internationalisation of logistics systems has a higher share among Chinese companies. On the other hand many of them follow a Greenfield approach when entering new markets, and the most successful ones (see Fig. 13) even apply quite sophisticated logistics planning. Hence, a clear trend can hardly be identified.

Logistics objectives for setting up shop in foreign markets. When entering a new market, various logistics objectives are thinkable. For this survey, two different types of objectives are distinguished. Firstly, objectives that are related to the internationalisation process itself, e.g. speed of implementing the logistics system, minimizing costs for setting up a new systems etc. Secondly, objectives that are related to the characteristics of the final system, e.g. reasonable operating costs, high reliability in delivery time etc.

Interestingly, the top three objectives of German companies are the same for all respective regions and focus all on the quality of the final system: minimize operating costs of the final system, high reliability in delivery lead time, and high process stability (see Tab. 3). Hence, the outcome of the internationalisation process is more important then speed and cost of the process itself. What counts, is the final logistics system.

In contrast, Chinese companies' objectives are more diverse and differ from region to region. A look at the Chinese preferences shows that the most important issue of the logistics strategy is the speed of setting up shop, followed by the operating costs of the final system, minimising the costs for setting up the new system, and finally high service levels with short lead times. The focus on speed and cost of setting up shop is also reflected in Fig. 14, where it could be shown that the internationalisation process for Chinese companies lasts significantly shorter than for their German counterparts.

Overall, a clear trend within the Chinese respondents' objectives can hardly be discovered. For each region a different set of objectives is most important and in total, the importance of each objective seems to be almost equally distributed. One possible

explanation could be the higher degree of
Muddling Through among Chinese logisticians
compared to their German counterparts, who
follow a limited set of strategic logistics objectives, regardless of the specific region.

Insight — Dynamic Developments in China's Automobile Industry

A significant example of the rapid Chinese economy development is the automobile industry. As China's reform period began in the late 1970s, the nation turned its attention to improving domestic car and truck production. By the year 2000, nearly every passenger car in Shanghai was made by a local Volkswagen (VW) factory, and Wuhan traffic was dominated by the city's Citroen venture's products. Beijing saw a large number of nearby Tianjin-made Daihatsu cars serve as the capital's low-cost taxi fleet. Overall, VW took some 60 % of the entire nation's passenger car market until just a few years ago.

Over the past five years, automobile production and sales in China have nearly tripled, and with more than 4 million vehicles made in 2003, the country now ranks with car powerhouses such as Germany and France, and is among the top-five auto makers in the world. The forecast for 2007 are 4.9 million sold automobiles in the Chinese market. Foreign corporations have played a major role to date in the explosive expansion of the automobile sector, but mainly through Sino-foreign joint ventures.

A spate of new foreign and domestic plant investments over the past two years has created a problem of looming production over capacity, with Nissan, Ford, Toyota and Hyundai joining Volkswagen, General Motors, Honda, Citroen and others in an increasingly crowded production arena. The automobile production was growing at a rate of 87 % in 2003. Thus imports of spare parts and components increased between 17.4 and 25.8 % during 2001 and 2005. The import of so-called Fully Built Up (FBU) vehicles to China for example, is highly restricted with duties. Today, only 5 % of newly registered cars are FBU in this country, while the content of local components of the Volkswagen model 'Santana 2000' has increased to 98 %. Hereafter a few figures of foreign car producers to underline the rapid growth in the automobile industry:

- Volkswagen AG invested € 5.3 bn in Changchun 2003 and also founded a new € 6 bn engine plant for 300,000 engines annually in Shanghai.
- Ford invested US$ 1 bn in China 2003 and plans to make further investment of US$ 16 bn until 2010.
- Delphi plans to develop injection pumps in China.
- Yamaha thinks about moving its R&D department division to China.

5.4 How to Cope with the Dynamic Environment

The first part of Chapter 5 deals with the internationalisation process, i.e. how it is organised, what steps it consists of, how logistics is being integrated in the decision making, and which objectives companies follow herein. The second part focuses on the issues that have to be considered when setting up international logistics systems.

Key components of logistics strategies for setting up offshore business entities.
No matter whether a company is engaged in strategic planning or acts according to the concept of incrementalism and regardless of what objectives the logistics strategy tries to accomplish, three key components have to be part of any logistics approach to ensure that the final logistics system will effectively support the foreign market entry: (a) the logistics system has to cope with rapidly changing conditions it will be operated in, (b) globally standardised logistics processes have to be balanced with local adaptation (comp. Chapter 5.5), and (c) decisions have to be made regarding the strategic collaboration with logistics service providers (comp. Chapter 5.6).

Rapidly changing business environment.
In order to obtain the full value of logistics as a source of competitive differentiation, managers evolve and implement logistics systems that are increasingly more complex, longer lasting, more difficult to reverse, and riskier than ever before. Ironically, the greatest risk may lie in failing to develop a logistics network responsive to the rapid pace of environmental change. As indicated in Chapter 4, Internationalisation increases dynamics and uncertainty in logistics networks to a large extent. The logistics conditions, e.g. infrastructure, legal regulations, administrative matters etc. are subject to constant change. Future development of markets and even the company's future success are — by its

nature — unpredictable. Considering that all these influencing factors change differently in each country makes clear, that the design and management of global logistics systems are challenging and error-prone tasks. Correspondingly, more then 80 % of the logistics managers stress that the dynamic changes of the business environment create the biggest challenge for internationalisation (see Fig. 20).

This overwhelming awareness of the fact that logistics systems have to cater unpredictable and rapidly changing requirements in the future, leads to the question of to what extent flexibility and agility is incorporated in the their planning and design.

Consideration of dynamics and uncertainty. More than 80 % of the reviewed companies insist that the AS-IS situation cannot form sufficient ground for planning logistics networks. During the planning and design of logistics capacities, location of facilities, network structure, definition of leadtimes etc., 80 % of the Chinese and 60 % of the German logisticians develop different scenarios, which are evaluated according to their future expectations. This is typically undertaken in cooperation with other corporate functions, to take the anticipations of other experts also into account, e.g. sales expectations.

Still, future developments are only estimated and will most likely be inaccurate. Hence, 72 % of the Chinese and 53 % of the German respondents focus on creating flexible logistics structures to cope with uncertainty.

Dynamics and uncertainty originate from various causes, e.g. shifting of customer's preferences over time, changes in legal restrictions, etc. In the early stages of China's economic upswing, for example, foreign investors located manufacturing facilities and distribution

Fig. 20 — Dynamic changes in global logistics

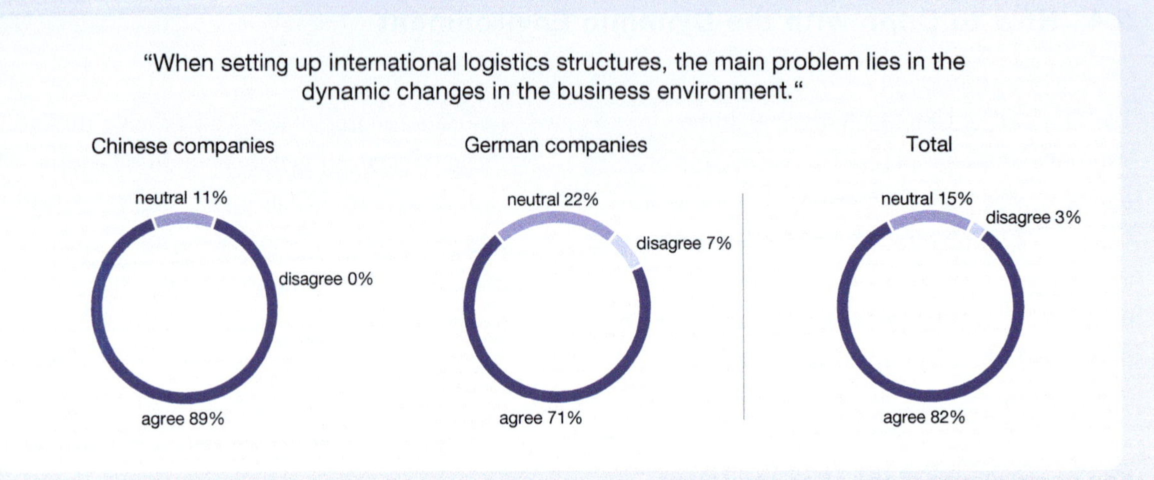

centres exclusively along the coastline. Nowadays, China's purchasing power is moving more and more into the hinterland and logistics structures with a focus on the coastal areas may not be efficient any more.

Regarding changes in the legal framework, one might think of China's WTO accession. For those who adapt to the resulting changes a multitude of opportunities arise. Those who do not immediately realise the transformation's consequences to their business might be quickly out of market.

In general, the dynamic and uncertain environment is perceived mostly in emerging markets. However, even in developed regions dynamics must not be underestimated. One prominent example is the eastern enlargement of the EU. When countries like Poland, Slovakia, and the Czech Republic joined the union, many companies felt that their centralised distribution structures in Western Europe had to be redesigned. Many central warehouses were removed eastwards, to cater the enlarged market.

Coping with dynamics and uncertainty.
Logistics flexibility has different dimensions, ranging from a system's ability to adapt to short-term changes in customer demand to the possibility to alter a network's structure according to major shifts in its geographic

focus. Various methodologies may be applied in accordance with a company's business model. That can be postponement strategies, agreements for flexible working hours, outsourcing concepts, etc. Logistics outsourcing is a hot topic since more than a decade and it becomes increasingly relevant in the context of foreign market entry (comp. Chapter 5.6).

More important than the exact arrangement of measures to raise flexibility, is the overall awareness of managers at strategic levels that the logistics system they set up in foreign markets has to be agile enough to fulfil tomorrows unpredictable business requirements.

Exit stategies for the case of market retreat. Managers often ground their positive expectations regarding international business expansion on the abundance of international success stories. But there are just as many famous examples of failure, where companies lost a fortune overseas and even decided to retreat from certain markets. The retreat of Wal-Mart from the German retail market is just one very recent example, another would be the German home improvement chain OBI, who decided to leave the Chinese market a few years ago.

The high degree of uncertainty regarding the market development leads to the challenge that logistics systems should be supportive

Fig. 21 — Incorporation of exit options

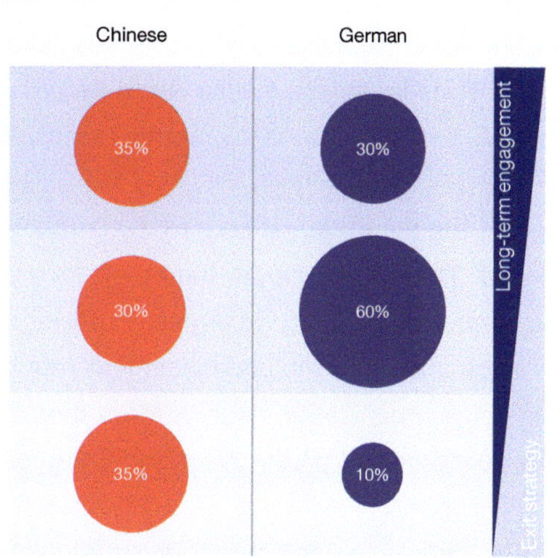

for a rapid expansion and, at the same time provide the possibility to retreat from a country with a minimum of sunk costs and residual overhead costs.

Even though the surveyed companies mostly reflect that when planning and designing international logistics systems they do consider that the environment of the system and its conditions might change at a fast pace, there can still be observed a poor consideration of failure in the new market.

Expectations concerning the success of internationalisation projects are very high. Hence, one third of the respondents regard any foreign market entry as a long-term engagement (see Fig. 21). 60 % of the German logisticians only include an exit option in their strategy, when they perceive the market particularly risky. Only 30 % of their Chinese peers show the same behaviour. Another 35 % of the Chinese and only 10 % of the German companies consider a failure in a foreign market always

possible, and consecutively install exit options in any new logistics system. Chinese companies seem to incorporate flexibility in their internationalisation approach better than German companies, which could be explained by the fact that companies in China are exposed to a much more dynamic environment than in Germany.

From a logistician's perspective, dynamics and uncertainty of the logistics environment can hardly be influenced. Hence, he has to deal with it. The degree to which flexibility is being incorporated into the set of objectives for foreign market entries still leaves room for improvement. The findings show that, in comparison with quantitative measures, e.g. costs and performance, flexibility receives relatively little attention. With regards to the justifiably high dynamics and uncertainty of today's business environment and the corresponding costs from inflexible logistics systems, this target system seems to be doubtful.

Insight — North America

In the understanding of this survey, North America comprises the United States of America (USA) and Canada. With only 6 % of the world population, North America ranks first with 33 % share of the world GDP. Its landmass makes North America the third largest continent on the globe with an area of almost 18.5 million square kilometres (simply USA and Canada).

The USA, the third biggest country by land area and population, is the strongest economic power of the world. Their GDP of more than US$ 13 trillion ranks first worldwide. In fact, the USA is the first largest importer and second largest exporter of goods. Additionally, the GDP grew at a 2.8 % annual pace since 2000 and the unemployment rate has fallen to only 4.6 % in 2006. The USA is also affected by a high urbanisation degree. More than 87 % of the population (302 million) live in urban areas, mainly concentrated on the West and East coast.

The USA have a convenient road system of interstate and U.S. highways, linking almost every urban agglomeration to an efficient road network. On the other hand the railroad system is comparatively scarce. The national rail organisation, Amtrak, provides services to many cities, but some parts, particularly the bridging of East and West, suffer from the poorly developed railroad tracks. In order to cross the continent from East to West, for instance, it can take up to an entire week. Because of this unfavourable duration many companies stick to the more expensive air transportation.

Consequently, the aviation sector in the U.S. has been of particular importance from the very first. Due to the large distances, air transportation is widely developed. The U.S. has the highest ratio of passenger to inhabitant world wide. In 2005, 660.9 million passengers were reported, which makes every inhabitant fly more than twice a year. Seven American airports are among the TOP 15 of the largest cargo airports in the world regarding cargo handling and even eight airports are listed in the TOP 15 of passenger capacity worldwide. Both lists are led by American airports with Memphis on top of the cargo airports and Atlanta as the number one in the passenger category. While FedEx, UPS, and United Airlines are the world leaders of air freight carriers in the category Revenue Tonne Kilometres (RTM), Delta Air Lines, American Airlines, and Southwest Airlines are in the vanguard of passenger transportation. In 2005, when the new aviation agreement drew to a close, the USA reached a full open skies aviation agreement with Canada. It allows an unrestricted number of flights by carriers between the countries and also eliminates restrictions of the kinds of aircraft flown, and prices charged.

The major harbours in the U.S. are in the northeast, responsible for freight originating from Europe and along the entire West Coast. The three most important harbours are located at Los Angeles, Long Beach, and New York responsible for almost 50 % of the container handling in the U.S. There are signs of fierce competition among West Coast ports from British Columbia to Southern California. The West Coast of North America is a logistics hub for Asian imports on a global scale, especially when China's rising began. China sends more than 12 million TEU to the U.S. each year. All that traffic has created an image for the ports that includes problems such as land constraints and infrastructure at or near capacity.

Canada with English and French as official languages is the largest country in North America by area and the second largest in the world after Russia. The population density of 3.5 inhabitants per square kilometre (33 million in total) is among the lowest in the world. Many rural areas are uninhabited. Canada is one of the wealthiest countries with a high income per capita of more than US$ 30,000. The

Canadian economy has been growing steadily over the last decades with a low unemployment rate at 6 %. The primary sector has an unusual important role in Canada compared to other developed countries such as in Western Europe. The oil and logging industries are among the largest. The country holds the second biggest oil and gas reserves in the world after Saudi Arabia which makes Canada a strong net exporter of energy. Canada is also the world leading producer of natural resources such as nickel, aluminium, zinc, uranium, and gold. However, the Canadian economy is dominated by service industry which employs more than 75 % of all Canadians. Canada is highly dependent on international trade, especially with its southern neighbour the U.S., which put the economy vulnerable to external fluctuations.

Due to the size of the country, it is hard to cover every area by road or railway. Nonetheless, the infrastructure in Canada is one of the most developed in the world. The very fact that the Canadian government made a historical investment of US$ 16 bn in 2007 in infrastructure for the next seven years is a sign of positive future developments. Canada is linking two sides of a coin, residential densities that look much like those in the U.S. and transportation characteristics that look more like in Europe.

Despite arguments about specific portions of the North American Free Trade Agreement (NAFTA), signed in January 1994, it is generally agreed that its effects have benefited the three signatories - the USA, Canada, and Mexico. The NAFTA has been in effect for 13 years. When it was founded in 1994, it represented a US$ 6 trillion economy with a population of 360 million. Within a decade the area grew to a US$ 12.5 trillion economy with a population of 430 million. The intention of the agreement was to dispose all trade barriers and tariffs among the three participating countries USA, Canada and Mexico. Restrictions were to be abolished in various fields, including computer applications, motor vehicles and textiles and. Intellectual Property Rights were also included.

Since the passage of NAFTA, the U.S. and Canada have experienced significant growth in trade, increased specialisation, and deepened integration of the two economies. Although the agreements liberalised trade and investment policies, they only partially facilitated a free flow of products across the border. However, in reality they have actually lead to an increase of processes and paperwork necessary for bilateral trade. These can be considered as an obstacle to the a free flow of transportation, and the resulting costs are in fact increasing logistics costs. The U.S. and Canada share the world's largest bilateral trading relationship. For 2006, total merchandise trade (export and import) between the two countries totalled US$ 533.7 bn. The U.S. is still Canada's most important market, and Canada is the largest single country trading partner of the United States. In 2006, total merchandise trade with Canada accounted for US$ 303.4 bn in imports and US$ 230.3 bn in exports. The U.S. and Canada are also major sources of FDI to each other.

The impact of globalisation and protecting business continuity are the two most important changes for the 3PL industry in North America. One of the major globalisation trends is an increasing number of U.S. and Canadian companies taking control of goods at their overseas point of origin before they are loaded into containers for export to these shores. This trend is particularly strong in the retail sector. In terms of business continuity, customers expect their 3PL to truly be a global player to ensure their business continues uninterrupted. Customers change their sourcing plans with increasing frequency, pushing their providers to quickly react and deliver the goods without any disruption to normal business operations. The last decade witnessed a significant shift in manufacturing production to low-cost countries, particularly China. China is the origin of almost 70 % of U.S. imports from Asia.

5.5 Global Standards and Local Adaptation

Besides the rising dynamics, another major challenge of global logistics networks is the inherent heterogeneity of processes. Caused by the huge number of involved locations, intermediaries, suppliers, and customers working with different methods and routines under distinct logistics conditions, global logistics management is an increasingly complex task (comp. Chapter 4).

A widely recognised approach to address complexity in global logistics networks is process standardisation. Similarly to the concept described in Chapter 5.3 about the attempt to standardise repetitive procedures of the internationalisation process itself, logistics activities in global logistics networks may be standardised as well. It means that complex logistics tasks are decomposed to a limited number of simple standardised process modules which are equally carried out at all locations, all over the world. Furthermore, complete logistics practices may be applied similarly at any place in the network, no matter whether it takes place in the domestic market of the company or overseas.

Potentials and characteristics of globally standardised processes. Global process standardisation can mitigate the problem of complexity in many ways. First of all, it enables companies to establish worldwide transparency by monitoring their networks with globally linked and unified IT systems. If all processes share a minimum of similarity a predefined set of KPIs may cover global supply chains in terms of quality, time and costs, making different parts of the network comparable. In case of disruptions or hold-ups it is easier to identify and solve relevant problems; process optimisation on a global level is facilitated as well as the implementation of best-practices at different locations. Furthermore, the exchange of staff members and managers

between different facilities becomes possible with a minimum of introduction. All these features of globally standardised processes enhance the supervision and management of logistics networks, reduce costs, save time and increase the reliability of operations.

For the individual company the range of processes to be standardised will differ according to their specific industry and business model. Basically all kinds of inbound, in-house and outbound processes could be subject to standardisation. One might think of boxes and pallets in use for the transportation and storage of goods, order cycles and replenishment strategies, the application of milk runs and KANBAN systems, production scheduling, Push-Pull strategies, the design and layout of cross docks and warehouses, the degree to which distribution networks are centralised or regionalised, the use of Auto ID technologies (e.g. bar codes and RFID), etc. Standardisation is by no means limited to a company's boundaries. It may also include partners, customers and suppliers in the supply chain as the collaborative processes and interfaces could be unified for enhanced cooperation.

Limitations of global standardisation. Despite its various beneficial aspects, the standardisation concept has limits. The local business environment in the different countries and regions, including the legal framework, infrastructure and customer expectations, call for local adaptation of logistics processes.

To give an example, the German retailer Metro has a unified concept for the design and layout of its cash-and-carry-outlets. The processes in the building are standardised. The replenishment, for instance, takes place by trucks that are unloaded at the back of the outlet and timeslots are defined, at what time of the day which segment of the assortment

Fig. 22 —   Balancing global standardisation with local adaptation

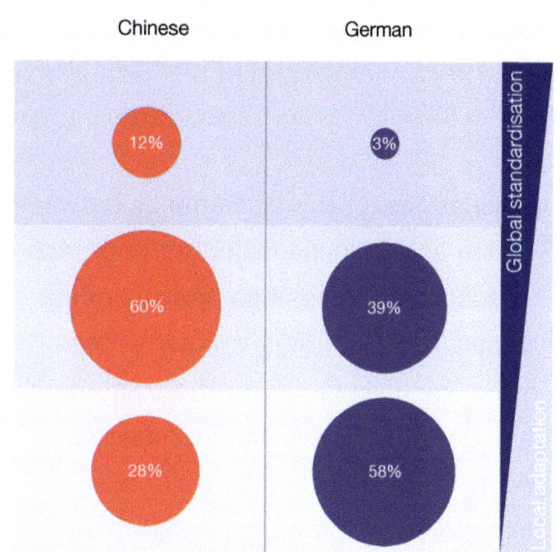

has to be delivered. Perishables have their specific timeslot, dry food has another etc. The standardised building concept and the respective processes have proven their practicability and efficiency in the huge number of Metro's outlets.

However, in different regions, customers have different habits and prefer different locations of stores. European consumers generally prefer stores that are situated near the city centres, while consumers in the U.S. like complex shopping centres at the outskirts. Again, consumers in China or India prefer shopping at corner shops. Different ways of shopping call for different types of stores and assortments. In many developing nations, for example, people shop on a daily or as-needed basis and carry bought items home, due to low incomes and a lack of storage and refrigeration space. As a result, goods need to be offered in small packages at a larger number of small shops handling only specific types of goods.

According to the peculiarities of the consumer behaviour and the market for real estate in Asia, Metro had to apply certain modifications on their standard concept for outlet buildings, to cater the local demand in those markets. This represents a typical example for the necessity to adapt to local conditions when companies go global.

Application of global standardisation versus local adaptation. The remarks and examples above lead to the question, to what extent companies apply standardised logistics processes when entering foreign markets, and how much adaptation it takes to meet the requirements of their target regions.

As illustrated in Fig. 22, 12 % of the Chinese and 3 % of the German respondents indicate to roll out their logistics system without any local adaptation. These companies simply transfer the design and structure of their domestic system to the foreign market, without modification according to local requirements and country-specific conditions. This may include modes of transportation, use of warehouses and cross docks, collaboration with service providers, and suppliers..

As in many cases specific local conditions prevent from a complete roll-out, the majority of Chinese companies (60 %) accept a minimum of local adaptation, but adheres to the idea to standardise logistics processes as much as possible. Some 39 % of the German respondents follow the same approach.

With 58 %, the majority of German companies develop completely new logistics systems for each and every market, to make sure it caters the specific local conditions as much as possible. About one quarter of their Chinese peers act similarly.

Two explanations for the discrepancy between the German and Chinese majority's approach may hold: at first, Chinese companies compete mostly in commodity markets, where low cost is the main factor for the customer's purchase decision. Therefore, standardisation seems to be adequate for them. German companies on the other hand, mostly offer products of higher sophistication, and feel that their logistics system has to provide superior customer service. Hence, German companies tend to develop country specific systems.

A second explanation could be drawn from the fact that Chinese companies, in general, apply a Greenfield approach when entering foreign markets (push strategy). This creates the opportunity to apply well designed and systematically planned standardised processes. German companies, seem to grow their networks more organically over time, often driven by business opportunities derived from customers' requests from overseas (market pull). Thereby, individual island solutions have evolved, that will later be optimised when necessary.

The most successful companies (see Fig. 13) apply higher levels of global standardisation than the average, and surprisingly, the German respondents do it even more than their Chinese peers.

Basically, the overwhelming majority of all respondents appreciate the benefits of a standardised international roll-out. The differences in the results come from the different perceptions to what degree each business model requires an individual country specific system. Legal and infrastructural necessities aside, one can say that the overall objective is to fulfil the customers' needs—but it has to be effectively balanced, to what extent he recognises and values regional adaptations and to what degree he would accept the global standard of the company.

Like in many other disciplines, the sentence "Think global, act local!" proves to be right, which means to have a global strategy in mind, and adapt it where necessary.

5.6 Strategic Collaboration with Logistics Service Providers

As indicated in Chapter 5.4, three components are key for any logistics approach in the context of foreign market entry: how to cope with the rapidly changing conditions in which the logistics network is operated (comp. Chapter 5.4), how to balance globally standardised logistics processes with local adaptation (comp. Chapter 5.5), and finally, how to strategically collaborate with logistics service providers.

Logistics outsourcing. In the developed economies, outsourcing of logistics activities has been a hot topic for decades, having boosted the continuously prospering industry of logistics service providers. Various types of LSPs have emerged, ranging from freight forwarders and 3PL to specialised logistics and SCM consultancies, offering a multitude of logistics services, including basic activities such as transportation, warehousing and handling, up to sophisticated and complex network solutions including information processing. Today's logistics revenues show the economic significance of the LSP industry; for Germany alone, the total logistics revenues in 2004 are estimated to be EUR 170 bn and for the region considered Western Europe in this survey (EU 15, Switzerland and Norway) calculations are about EUR 729 bn. For China, estimations for the value of total logistics services (including outsourced services and self-run logistics activities) vary between EUR 250 bn and EUR 320 bn, which is roughly one-fifth of the Chinese GDP.

In the context of internationalisation and the inherent dislocation of value adding activities, the outsourcing of logistics activities becomes even more important as compared to a mainly national environment. Hence, manufacturers and retailers increasingly rely on logistics service providers when setting up shop in foreign markets.

Importance of logistics service providers for foreign market entry. Less than one-third of the reviewed companies in this survey indicated, that they need to have self-run logistics assets and operations to maintain the right quality and service levels for their business. Though, most of these only keep a limited range of specific operations which they perceive to be most important under their control. Very few manufacturing and retail companies own and manage the complete range of their logistics activities.

Companies entering foreign markets outsource an even broader range of activities when going global in order to reduce the risk of the market entry and to leverage the country specific know-how of their partner. Clearly, it is now a strategic decision to outsource logistics activities; in addition, the role of LSPs has changed as they have been allowed to penetrate further into their customers' operations and supply chains.

In case that in the new target market, there is no adequate and capable LSP available, 71 % of the Chinese and 61 % of the German managers would rather help a new LSP to develop in that market, than investing in dedicated self-run assets. This clearly shows the strong coherence of foreign market entry and global logistics outsourcing.

Hence, the LSP industry should address this strong willingness to outsource, by providing services that further reduce market entry barriers for their customers. Many of them do not have the knowledge and expertise to carry out logistics operations in foreign countries by themselves because until recently, in most cases, their logistics activities have been confined to the countries or regions in which they

Fig. 23 — Objectives of logistics outsourcing when going global

are located. Especially services that support the global expansion of SMEs have a huge potential.

Strategic objectives of logistics outsourcing. As mentioned before, the time aspect is of major strategic importance for successful market entry (comp. Chapter 5.1). Consequently, the highest ranked objective for the collaboration with LSP is to ensure a quick market entry by leveraging the LSP's existing global network (for the complete sample, see Fig. 23).

For German logisticians, the LSP's cost advantages from his ability to benefit from economies of scale are most important.

As setting up global logistics networks requires regional expertise in order to cope with local culture, politics, and environmental pe-

culiarities, Chinese managers value the LSP's specific know-how and service quality as the strongest argument for outsourcing. As LSPs typically have already tremendous local market expertise and knowledge in the respective country or region, and are familiar with the legal and administrative regulations, in addition to possessing the required licenses etc., the market entry is significantly facilitated.

Another important aim of outsourcing is to keep investments for the market entry low and thereby minimise the inherent risks with asset-light strategies. Furthermore, collaboration with LSPs may improve the logistics flexibility of a company. In general, LSPs have better possibilities to adjust capacities in short term than their clients themselves. As they serve multiple customers, the volatility in the demand of each individual company is very likely to be balanced out by the aggrega-

Fig. 24 — Outsourcing strategies and composition of self-run logistics

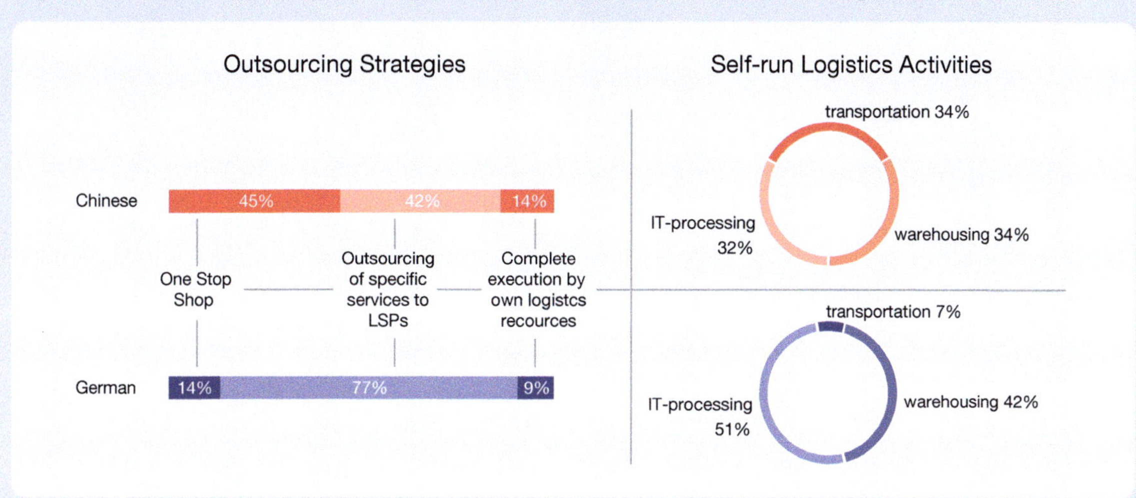

tion at the LSP. Additionally, 76 % of the Chinese and 48 % of the German respondents try to raise their flexibility by enforcing short contract periods.

Outsourcing levels and self-run logistics activities. While Chinese and German companies both appreciate the benefits from logistics outsourcing, their concrete collaboration with LSP shows certain distinctions. In general, Chinese companies show significantly lower outsourcing levels then their German peers (see Insight — China). Though, Chinese upcoming global players who are now entering foreign markets show a clear tendency towards higher outsourcing levels.

Regarding the actual cooperation with LSP, 45 % of the Chinese respondents prefer one-stop-shopping, i.e. they work with only one partner who provides a global scope and can cover all needed services. Oftentimes the service provider reaches geographical coverage and the full range of activities only by close cooperation with further partners and subcontractors. But from the customer's perspective the advantage is to have only one partner to be in touch with, hence, a simplified coordination.

However, only 14 % of the German respondents demand one-stop-shop solutions. Know-

ing, that the LSP cannot provide all required services and therefore has to work with subcontractors leads to certain scepticism, whether it is ensured that all partners and intermediaries involved are sufficiently reliable and how they commit to the defined rules and quality standards. Furthermore, manufacturers and retailers are still restraining to rely on one provider only, because they do not want their global logistics procedures to depend on the performance of one single third party. Besides arising problems in case of disruptions, bargaining power and switching costs are further strong arguments not to rely on a single LSP.

With 72 % of the German and 42 % of the Chinese, the clear majority of logistics managers outsource specific activities to several LSPs. The overall control of the logistics and supply chain network remains at the customer side, and the limited range of services per provider ensures not to become too dependent on certain LSPs.

Looking at the range of outsourced logistics activities, customers often ask for value added services and global capabilities, but these categories are, as of yet, less important for the final outsourcing decision. Moreover, logistics outsourcing is still based on traditional services and to a lesser degree on a more holistic approach where the full supply chain

Tab. 4 — Criteria of LSP-selection

Top three criteria for selecting LSP*	Eastern Europe	Russia	South America	North America	India	China	Western Europe
Price	▲	• ▲	• ▲	• ▲	• ▲	▲	
Logistics quality, reliability	• ▲	• ▲	• ▲	• ▲	• ▲	▲	•
Experience gained from former cooperation	▲	▲	▲	▲	▲	▲	
Reliability/Process Quality				•	•		
Motivation and quality of the management in charge							•
Flexibility and adaptability	•	•					
Reputation, brand, image		•	•	•			•
Viability/Future development potential			•				
Size/Revenue	•		•				•
Technology equipment and skills							•
Endowment with capital for future investment in charge							

Importance

* When unable to specify the top 3 categories due to equal numbers of response, all items with the top three frequencies are stated.

• Chinese ▲ German

is outsourced, mainly because of customers' concerns. These common concerns are the potential loss of direct control of logistics activities, uncertainties about the service levels provided by the outside company, and questions concerning the true costs of employing a third party.

Only 14 % of the Chinese and 9 % of the German reviewed companies completely execute their logistics activities by themselves with self-owned assets and facilities.

The composition of activities that companies do not fully outsource when going global, also shows strong differences between German and Chinese companies. Among the Chinese respondents, transportation, warehousing, and IT processing, all show an equal likelihood not to be outsourced. German respondents indicate that self-run transportation is very exceptional. Of German companies, 42 % would have self-run warehousing in foreign markets and — with the largest share — 52 % of the German respondents tend to keep the IT processing of their global logistics networks in-house (see Fig. 24).

Criteria for LSP selection. Having made decisions about the extent to which logistics activities should be outsourced and whether the complete volume should be handled by one or more partners, it is time for the LSP selection.

The survey determines the most important selection criteria for LSP when setting up shop in foreign markets (see Tab. 4). First of all, it shows that the key factor nowadays is price, while the next-highest factor rated is the quality and reliability of the services provided. Both are not mutually exclusive but must be balanced in particular. When companies focus on price as criterion, it is important to keep a holistic view on the overall costs. Customers should be aware that a fruitful collaboration with a capable LSP may significantly improve the way in which products are moved around the world – and thereby reduces costs in consequence. This potential for savings throughout the global logistics network may by far outweigh the direct price for the LSP's services, because logistics efficiency can then be improved by considering other issues such as quality, operations, and technology.

According to the complete sample of respondents, the third most important criterion to select a LSP for foreign market entry is the experience gained from former cooperation. For Chinese companies, this even is the most important criterion of all. One-third of the Chinese and 10 % of the German logisticians indicate that they would not go into a new market without the LSP they work with in their domestic region. Almost 50 % of the Chinese respondents always work with one of the renowned global LSP, only one-third of the German logisticians do it the same way.

Analysing the set of criteria according to the different target regions and countries, it shows that German companies do not change their preferences with regards to the regions. Their top 3 criteria are the same for all. In contrast, the Chinese respondents make certain distinctions, e.g. in Eastern Europe and Russia, flexibility is of comparatively higher importance, in South America they care relatively more about the viability and future development potential of their partners.

The observation that German companies in this respect show a clear pattern in their selection criteria while Chinese companies' criteria are more equally distributed and vary more strongly according to the target regions, might be another evidence that Chinese companies follow the Muddling Through concept more than their German peers.

Global logistics industry. Overall, the results of the survey show that through their global expansion, companies have heightened expectations regarding logistics outsourcing. They increasingly require complex and specialized logistics solutions from their LSP who have to continuously improve their capabilities and facilities. Accordingly, the competition in the logistics industry has sharpened so that logistics service contracts become ever more sophisticated and place more pressure on providers to continually improve processes.

The results clearly show that, beside large operational networks with huge geographic coverage, customers also appreciate the LSP's local logistics expertise of each region and country. However, despite the existence of major global players in the logistics markets and huge M&A activities in recent years, a single provider with the level of knowledge and infrastructure required to truly operate globally without subcontracting and cooperation, can currently hardly be identified. In fact, only a few of the LSPs, because of their size and scope, can ensure that a project is properly implemented worldwide, and that their own people will be working directly with customers in all locations.

6 Conclusion

The survey provides clear evidence for the superior significance of logistics for the successful internationalisation of business activities.

The process model developed for this research has enabled the conceptual analysis of how companies set up shop in foreign markets and what role logistics plays herein. The results of the most successful Chinese and German companies show, that the intense integration of logistics issues and the early integration of logistics managers into the decision making process and the business design of foreign market entries are key success factors.

Though, a standard logistics blueprint for internationalisation which guaranties success cannot be identified. The different business models, strategies, and market peculiarities make it necessary for every company to find the logistics concept that matches best with its individual needs. On average, German companies tend to apply a more formalised approach with a higher degree of central strategic planning. Among Chinese companies a higher share of respondents follow a Muddling Through approach. But the consolidated data of the most successful respondents paints a different picture — here, the most successful Chinese companies show advanced logistics planning and in many cases analytical global footprint design.

Comparing German and Chinese companies' internationalisation strategies it has to be understood that they have completely different backgrounds and starting positions. Most German companies have grown their international logistics networks more organically over time, often driven by business opportunities derived from customers' requests from overseas. Step by step, and oftentimes without a clear global strategy individual island solutions have emerged that are being optimized when necessary. Currently, these companies are more experienced in foreign market entries than their Chinese counterparts. On the other hand, only a limited number of Chinese companies are yet ready to begin their global expansion. Typically, these companies apply a Greenfield approach when entering foreign markets, which creates the opportunity to conduct in depth analysis of market developments and conditions and to apply a holistic global planning for global footprint design.

The level of logistics sophistication that is observed at the Chinese respondents' answers might be surprising to those who have already gained experiences with Chinese companies in China. But it has to be stressed, that the sample does not represent the average Chinese firm, which is mostly competing in the domestic market. Only those companies that already run international operations or are about to do so have found their way into this survey. Hence, they represent the avantgarde of Chinese firms. They are typically well endowed with capital, well-educated talent, skilled leadership, and oftentimes employ managers with overseas experience. Hence, these upcoming Chinese global players may not be underestimated. They are well positioned to cause major changes in the global market place.

Considering that the overall importance of logistics and supply chain management for corporate success has increased dramatically over the past decades, it is clear that it is even more important to incorporate these issues in internationalisation activities. Three key components have been identified that are mandatory to be integrated in any global logistics approach. Firstly, any global logistics network has to be designed with respect to the fact that the conditions under which it will be operated will change at fast pace. And the business requirements it hast to meet are subject to constant change. Hence, high levels of flexibility and agility have to be ensured.

Secondly, as international logistics networks are connecting increasingly diverse regions and markets, logisticians have to deal with high levels of complexity. Therefore global standardised processes have to be balanced with local adaptation.

The third aspect is the strategic collaboration with LSPs. Its benefits are widely recognised and it shows that the outsourcing levels as well as the range of activities outsourced increase in the context of internationalisation. This is observed at both Chinese and German companies.

The conditions for the establishment of international business activities and global logistics networks have probably never been better than today. Though, it is not an easy task. It remains challenging for managers from various fields, especially for logisticians. As the topic is so diverse and complex, it is not surprising that some of the respondents' statements might even appear somewhat contradictory. Internationalisation is not a standard procedure and the related issues are by no means completely clarified. In practice, a lot of issues have to be solved with gut feeling and pragmatism.

Being so rich in topics the fascinating field of international logistics management can by no means be sufficiently covered by this survey. A multitude of questions remain unanswered and provide the fruitful ground for future research in the years to come.

References — Research Report

Baumgarten, H.; Darkow, I.-L.; Zadek, H. (ed.) (2004): Supply Chain Steuerung und Services: Logistik-Dienstleister managen globale Netzwerke – Best Practices; Published by Springer Verlag; Berlin et al. 2004.

Baumgarten, H.; Herter, M. (1999): Internationalisierung der Logistik; in: Weber, J.; Baumgarten, H. (ed.): Handbuch Logistik 1999 – Management von Material- und Warenflußprozessen; Published by Schäffer Poeschel Verlag; Stuttgart 1999; pp. 828-41.

Beatson, D. (2004): Becoming Truly Global: 3PLs face new Challenges; Source: http://www.inbound-logistics.com/articles/3plline/3plline0104.shtml; Access on March 11, 2007 at 11:37 p.m.

Bowman, R.J. (2006): Globalization, Customer Demand Drive Logistics Outsourcing to the Next Level; in: Global Logistics and Supply Chain Strategies; Vol. 10; No. 6; June 2006; pp. 40-5.

Busse, M. (2002): Transaktionskosten und Wettbewerbspolitik (Transaction Costs and Competition Policy); in: Wirtschaft und Wettbewerb, Vol. 52, Iss. 2, pp. 112-120.

Butts, J. (2006): The Seven Abilities of Highly Effective 3PLs; Source: http://www.inboundlogistics.com/articles/3plline/3plline0906.shtml; Access on March 11, 2007 at 16:02 p.m.

Capgemini; Langley, C.J.; SAP; DHL (ed.) (2006): 2006 Third-Party Logistics – Results and Findings of the 11th Annual Study; Georgia Institute of Technology; Atlanta 2006.

DHL (ed.) (2005): Logistik-Lotse 2005. Leifaden für Logistik – eCommerce – KEP – Spedition – Land-, Luft- und Seeverkehre – Zoll – Außenwirtschaft; 13th, revised edition; Published by Verkehrs-Verlag J. Fischer; Dusseldorf 2005.

Foster, T.A. (2005): Logistics Inside China: The Next Big Supply Chain Challenge; in: Global Logistics and Supply Chain Strategies; Vol. 9; No. 9; September 2005; pp. 28-36.

Foster, T.A.; Armstrong, R. (2006): The Top 25 Global 3PLs: Is Bigger Really Better?; Source: http://www.glscs.com/archives/05.06.25_ 3pls.htm?adcode=90; Access on December 14, 2006 at 17:00 p.m.

Germain, C.; Andrews, S.; Vera, H.; Dunsmore, S. (2006): Worth Repeating – Four Experts outline the Challenges of Global Trade Management; in: Canadian Transportation & Logistics; March 2006; p. 42.

Gershenhorn, A. (2004): The Making of a Successful Global Supply Chain; in: World Trade; Vol. 17; No. 12; December 2004; pp. 19-22.

Gourdin, K.N. (2006): Global Logistics Management: a Competitive Advantage for the 21st Century; 2nd edition; Published by Blackwell Publishing; Oxford 2006.

Harps, L.H. (2003): Global Logistics: Bridging the Culture Divide; Source: http://www.inboundlogistics.com/articles/features/0303_feature01.shtml; Access on December 10, 2006 at 13:30 p.m.

Hoffman, K.C. (2002): Global Growth of 3PLs Is in Response to Customer Demand; Source: http://www.glscs.com/archives/03.02.outsource.htm?adcode=90; Access on December 12, 2006 at 20:15 p.m.

Hofstede, G. (2005): Cultures and Organizations: Software of the Mind; revised and expanded 2nd edition; Published by McGraw-Hill; New York 2005.

Kerr, J. (2006): What's the Right Role for Global 3PLs?; in: Logistics Management; February 2006; pp. 51-4.

Lee, H. L. (2004): The triple-A supply chain; in: Harvard Business Review, Vol. 82 No. 10; pp. 102-112.

Loderhose, B. (2007): Frische Fernöstlich; in: Lebensmittel Zeitung; Vol. 20; May 18, 2007; pp. 32-33.

Ma, S.; Lin, Y; Chen, Z. (2000): Supply Chain Management. China Machine Press, 2000.

MacDonald, A. (2006): Managing a Global vs. Domestic Supply Chain; in: World Trade; Vol. 19; No. 7; July 2006; http://ctl.mit.edu/public/070606_WorldTrade_Sheffi_GlobalvDom.pdf; Access on November 30, 2006 at 21:34 p.m.

Mentzer, J.T.; Myers, M.B.; Stank, T.P. (2006): Handbook of Global Supply Chain Management; Published by Sage Publications; London 2006.

Mercer Management Consulting (ed.) (2005): Mythen der Kontraktlogistik: Mercer Studie untersucht Logistik-Geschäftsmodelle; Source: http://www.einkauf-und-management.at/index.php/einkauf/more/mythen_der_kontraktlogistik; Access on November 10, 2006 at 16:31 p.m.

Nohlen, D. (ed.) (1998): Lexikon Dritte Welt – Länder, Organisationen, Theorien, Begriffe, Personen; completely revised new edition; Published by Rowohlt Taschenbuch Verlag; Reinbek 1998.

Pfohl, H.-C. (2004): Logistiksysteme – Betriebswirtschaftliche Grundlagen; 7th, corrected and updated edition; Published by Springer Verlag; Berlin et al. 2004.

Piszczalski, M. (2002): Global Logistics Issues; in AD&P; January 2002; pp. 16-8.

Richardson, H.L. (2005): How to maximize the Potential of 3PLs; in: Logistics Today; May 2005; pp. 19-21.

Simchi-Levi, D.; Kaminsky, P.; Simchi-Levi, E. (2003): Designing and Managing the Supply Chain: Concepts, Strategies, and Case Studies; Published by McGraw-Hill; 2nd edition; New York 2003.

Straube, F. (2004): e-Logistik – Ganzheitliches Logistikmanagement; Published by Springer Verlag; Berlin et al. 2004.

Straube, F.; Dangelmaier, W.; Günthner, W.A.; Pfohl, H.-C. (2005): Trends und Strategien in der Logistik – Ein Blick auf die Agenda des Logistik-Managements 2010; in: Bundesvereinigung Logistik e.V. (ed.); Published by Deutscher Verkehrs-Verlag; Hamburg 2005.

Transport Intelligence (ed.) (2006): Global Contract Logistics 2006: Outsourcing and Collaboration; Brinkworth, United Kingdom; May 2006.

Tufinkgi, P. (2006): Logistik im Kontext internationaler Katastrophenhilfe; in: Schriftenreihe Logistik der Kühne-Stiftung 9; Published by Haupt Verlag; Bern 2006.

Vijayaraghavan, T. A. S. (2001): Impact of Transportation Infrastructure on Logistics in India; Source: http://www.findarticles.com/p/articles/mi_qa3766/is_200101/ai_n8952924; Access on January 4, 2007 at 18:44 p.m.

Waters, D. (2007): Global Logistics: New Directions in Supply Chain Management, Published by Kogan Page; London 2007.

References — Excursions

AAPA World Port Rankings 2005; Source: http://aapa.files.cms-plus.com/Statistics/WORLD%20 PORT%20RANKINGS%202005.xls; Access on August 30, 2007 at 03:40 p.m.

Auswärtiges Amt (2007): Indien: Wirtschaft; Source: http://www.auswaertiges-amt.de/diplo/de/Laenderinformationen/Indien/Wirtschaft.html; Access on June 7, 2007 at 02:45 p.m.

Auswärtiges Amt (2007): Länderinformation: Russische Föderation; Source: http://www.auswaertiges-amt.de/diplo/de/Laender/RussischeFoederation.html; Access on May 2, 2007 at 1:00 p.m.

Auswärtiges Amt (2007): Wirtschaftsdatenblatt Indien; Source: http://www.auswaertiges-amt.de/diplo/de/Laenderinformationen/Indien/Wirtschaftsdatenblatt.html; access on June 5, 2007 11:10 a.m.

Biedermann, D.; Dibenedetto, B. (2007): Twist of Fate; in: Journal of Commerce; New York; Apr. 9, 2007; p. 1.

Bundesagentur für Aussenwirtschaft (2006): Russlands Logistiksektor kann Nachfrage kaum decken; Source: http://www.bfai.de; Access on August 30, 2007 03:20 p.m.

Bundesagentur für Aussenwirtschaft (2006): Transport und Logistk – Russland; Source: http://www.bfai.de; Access on May 03, 2007 02:20 p.m.

Charles, A. (2007): The China Business Handbook 2007; 10th, completely revised and updated edition; Published by Alain Charles Publishing; London 2007.

Chiarello, A. A. (2005): Maersk Logistics USA Inc.; in: Journal of Commerce; New York; Jan 12, 2004; p. 1.

Cygnus Business Consulting and Research (2006): Background note – Logistics; Source: http://www.cygnusindia.com/articles/White%20paper%20on%20CII%20conf%20on%20Logistics%20-Kolkata%2080906.pdf; Access on 30 August, 2007 at 05:05 p.m.

Davies, Tim (2002): Argentina: Crisis? What Crisis?; in: Credit Management; Oct 2002; pp. 35-36.

de Schmidt, A. (2007): Provider am Ganges; in: Logistik Heute 05/2007; pp. 54-55.

de Schmidt, A. (2007): Take off am Taj Mahal; in: Logistik Heute 05/2007; pp. 48-49.

de Schmidt, A. (2007): Zäher Prozess; in: Logistik lleute 05/2007; pp. 56-57.

Dibenedetto, Bill (2007): Growing Pains; Journal of Commerce; New York: May7, 2007; p. 1.

Energy Information Administration (2007): Country Analysis Briefs: Russia; Source: http://www.eia.doe.gov/emeu/cabs/Russia/Background.html; Access on May 03, 2007 at 10:00 a.m.

Federation of Indian Chambers of Commerce and Industry (2007): Current State of Indian Economy; Source: http://www.indiainbusiness.nic.in/indian-economy.pdf; Access on August 30, 2007 at 05:00 p.m.

Franz, M.; Hassler, M. (2007): Eine Milliarde Kunden; in: Logistik Heute 05/2007; pp. 58-60.

Gakenheimer, R. (1997): The urban transport crisis in Europe and North America; in: Journal of American Planning association; Vol. 63; Iss. 4; pp. 530-531.

Gallagher, J. (2007): Racing for NAFTA freights; in: Traffic World, Newark; Mar 12, 2007; p. 1.

Göpfert, I. (ed.) (2006): Logistik der Zukunft – Logistics for the Future; 4th, updated and revised edition; Published by Gabler Verlag; Wiesbaden 2006.

Hausman, W.H.; Lee, H.L.; Subramanian, U. (2005): Global Logistics Indicators, Supply Chain Metrics, and Bilateral Trade Patterns; Source: http://rru.worldbank.org/Documents/Discussions/global_logistics_indicators.pdf; Access on November 22, 2006 at 18:12 p.m.

Hemerling, J.; Michael, D. C.; Michaelis, H. (2006): China's Global Challengers: The Strategic Implications of Chinese Outbound M&A; Published by Boston Consulting Group; May 2006; Source: http://www.bcg.com/publications/files/Chinas_Global_Challengers_May06.pdf; Access on 03 September at 04:45 p.m.

Invest in Germany (2005): Germany: Europe's Logistics Hub – Strengths and Opportunities; Source: http://www.kompetenzcluster.org/fileadmin/vdidaten/Logistik/Fakten/Germany-Logistics-Hub.pdf; Access on August 30, 2007 at 03:35 p.m.

Jahns, C.; Darkow, I.-L.; Weigl, T. (2006): Dynamic Supply Chains in Russia: Industries, Strategies, and Logistics Structures; Published by Bundesvereinigung Logistik e.V. (German Logistics Association); Deutscher Verkehrs-Verlag, Hamburg 2006.

JSC Russian Railways Company Profile (2007); Source: http://www.eng.rzd.ru/page.html?nav_id=145; Access on August 30, 2007 03:30 p.m.

Klaus, P.; Kille, C. (2006): Die TOP 100 der Logistik; 4th edition; Published by Bundesvereinigung Logistik e.V. (German Logistics Association), Deutsche Verkehrs-Zeitung; Deutscher Verkehrs-Verlag, Hamburg 2006.

Lopez-Carlos, A. et al. (2006): The Global Competitiveness Index: Identifying the Key Elements of Sustainable Growth; Source: http://www.weforum.org/fweblive/groups/public/documents/wef_member_pdf/gcr_0607_1_1_gcindexes.pdf; Access on August 30, 2007 at 1:00 p.m.

Morton, R. (2006) NAFTA: Twelve years after; in: Logistics Today; Vol. 47, Iss. 2; p. 10.

Open Joint Stock Company Russian Railways (2006): Annual Report 2005; Moscow (2006), Source: http://www.eng.rzd.ru/images/download_im.html?id=3810; Access on August 30, 2007 at 03:45 p.m.

Schulz, J.D. (2006): DHL crashes the party; in: Logistics Management; Vol. 44; Iss. 8; pp. 59-62.

Sissel, K. (1995): Foreign investment picks up; in: Chemical Week. New York; Nov 15, 1995; pp. 30-31.

Smith, James L. (2005): Calculating investment potential in South America; in: World Oil; June 2005; Vol. 216; Iss. 6; pp. 117-221.

Staudenmayer, M. (2007): Hohe Qualität in kleiner Stückzahl; in: Logistik Heute 05/2007; pp. 50-51.

Taylor, J.C., Robideaux, D.R., Jackson, G.C. (2004): U.S.-Canada Transportation and Logistics: Border Impacts and Costs, Causes, and Possible Solutions; in: Transportation Journal; Lock Haven; Fall 2004; Vol. 43, Iss. 4; pp. 5-16.

Verstaen, J.; de Schmidt, A. (2007): Achillesferse mit Schlaglöchern; in: Logistik Heute 05/2007; pp. 52-53.

Waters, D. (2007): Supply Chain Risk Management: Vulnerability and Resilience in Logistics, Published by Kogan Page; London 2007.

World Bank (2005): From Transition To Development: A Country Economic Memorandum for the Russian Federation; Publihed by ABSOLUT Print, Moscow 2005; Source: http://ns.worldbank.org.ru/files/cem/CEM_final_eng.pdf; Access on August 30, 2007 03:00 p.m.

World Bank (2006): Decentralization in India; Source: http://go.worldbank.org/YQWVRQ7NO0; Access on May 05, 2007 03:20 p.m.

World Bank (2006): Governance in India; Source: http://go.worldbank.org/JZ4EGDJQ30; Access on May 03, 2007 03:20 p.m.

World Bank (2007): Growth in India; Source: http://go.worldbank.org/1DLEBZ7C50; Access on May 04, 2007 02:40 p.m.

World Bank (2007): India at a glance; Source: http://devdata.worldbank.org/AAG/ind_aag.pdf; Access on 30 August, 2007 at 05:00 p.m.

World Bank (2007): India Foreign Trade Policy; Source: http://go.worldbank.org/RJEB2JGTC0; Access on August 30, 2007 at 04:30 p.m.

World Bank (2007): India Transport Sector; Source: http://go.worldbank.org/FUE8JM6E40; Access on August 30, 2007 at 04:30 p.m.

World Trade Organization (2007): Accessions: Russian Federation; Source: http://www.wto.org/english/thewto_e/acc_e/a1_russie_e.htm; Access on August 30, 2007 03:00 p.m.

WTO (2006): WTO International Trade Statistics, Source: http://www.wto.org/english/res_e/statis_e/its2006_e/its06_toc_e.htm; Access on August 30, 2007 at 03:55 p.m.

Frank Straube · Shihua Ma · Michael Bohn

Internationalisation of Logistics Systems

Frank Straube · Shihua Ma · Michael Bohn

Internationalisation of Logistics Systems

How Chinese and German companies
enter foreign markets

Springer

Frank Straube
TU Berlin
Strasse des 17. Juni 135
10623 Berlin
Germany
straube@logistik.tu-berlin.de

Michael Bohn
TU Berlin
Strasse des 17. Juni 135
10623 Berlin
Germany
bohn@logistik.tu-berlin.de

Shihua Ma
Huazhong University of Science & Technology
Wuhan
430074 Hubei
P.R. China
shihuama@public.wh.hb.cn

ISBN 978-3-540-76982-8   e-ISBN 978-3-540-76984-2

DOI 10.1007/978-3-540-76984-2

Library of Congress Control Number: 2007939801

Cover Design: WMXDesign GmbH, Heidelberg

Printed on acid-free paper

9 8 7 6 5 4 3 2 1

springer.com

德国国际物流网络能力中心，柏林

　　德国柏林工业大学物流系(TUB)在2005年组建了国际物流网络能力中心，由瑞士
Kuehne基金会提供研究经费。它的目标是同世界各国的高等院校和企业的合作伙伴研
究国际物流系统。柏林技术管理学院院长、柏林工业大学物流系主任Frank Straube教
授与Helmut Baumgarten教授任物流网络能力中心主任。他是德国物流协会(BVL)的副
主席、欧洲物流协会(ELA)的董事。他积极参与欧洲物流行业的活动，出版了许多关于
物流系统的专著，同时他也是国内国际会议上发表演讲的知名人士。Michael Bohn获
得汉堡和新加坡的经济工程硕士学位。他在制造和咨询行业富有经验。2005年任职于德
国柏林工业大学物流系。在国际物流网络能力中心他是关于国际物流网络的研究协作伙
伴。他主管国际物流范围的研究和国际研究项目。
详见: www.internationalelogistik.de

供应链与物流管理研究所，武汉，中国

　　1997在中国武汉华中科技大学(HUST)成立了供应链与物流管理研究所(ISC&LM)。
它是最早成立的研究物流与供应链管理的研究所之一。近年来它与许多海外大学进行了
广泛紧密的合作。武汉华中科技大学管理学院副院长马士华教授任该所主任。他长期从
事物流与供应链管理的研究与实际应用，参与生产与企业管理。他在该院为学士、硕士
生讲授供应链管理、企业管理、物流管理课程。他有丰富的为许多在中国的公司提供咨
询服务经验。
详见: www.logistics-chain.com

作者

柏林工业大学物流系主任
Straube教授

华中科技大学，武汉
管理学院副院长　马士华

柏林工业大学物流系
研究协作伙伴，Bohn硕士

研究协作伙伴

柏林工业大学物流系
Rief硕士

柏林工业大学物流系
Pohl硕士

华中科技大学，武汉
李昆鹏博士

在读大学生

- Roman Figiel
- Sebastian Flucht
- Pravin Narkhedkar
- Jia Qu
- Christopher Schwarz
- Kuo-Hsiang Yong
- 张欣
- 黄焜
- 李慧君
- 辛晶
- 魏晨
- 关旭
- 彭秉宗
- 雷征
- 黄铮

编者

- Axel Haas

尊敬的读者：

　　如今几乎任何一个公司都是全球供应链网络中的一部分，这个供应链网络极其依赖高效可靠的物流系统。对于任何规模的企业而言，全球化都为其创造了大量进入新的区域市场的机会和需求，对大量其他的参与者来说也是如此。这些构成了"物流系统的国际化"的研究核心。这项研究能够为德国和中国的合作带来大量机会。中国是全球最大的出口国，是近年来经济发展最快的国家，本次调查主要针对的也是这两个国家。

　　本次调查确定了中国企业和德国企业在国际化扩张中所采用的物流战略。调查涵盖了国际化活动中七个最重要的目标地区。从物流的角度确定了不同市场特殊的挑战和机会，由此比较了中国企业和德国企业是如何设法取得成功的。

　　此项研究由德国柏林工业大学（TUB）物流系国际物流网络能力中心与华中科技大学（HUST）管理学院供应链与物流管理研究所共同完成。能力中心由瑞士的Kuehne基金提供经费。这样的组合也体现出本调查具有极高的国际合作背景。

　　此次中德的研究合作从一开始就卓有成效。双方于2005年一月在武汉第一次会面，启动了项目。接下来的问卷设计和对结果的全面分析都是在非常友好的氛围下，经过大量的协作和坦诚的思想交流而完成的。

　　本报告旨在满足企业在全球物流网络中日常运作的需要，同时也能够满足学术研究者对物流全球化现象的兴趣。我们希望此次调查对加深物流国际化的理解有所贡献，对那些构建未来全球物流网络和相应的物流从业人员有一定的帮助。

　　我们真诚地感谢所有在调查过程中给与支持的企业、物流主管和研究者，还有那些接受访问和填写调查问卷的人员。我们诚心感谢你们的支持。

　　来自柏林和武汉的问候：

柏林工业大学物流系主任
Straube教授

华中科技大学，武汉
管理学院副院长　马士华

柏林工业大学物流系
研究协作伙伴，Bohn硕士

背景与调查总结

在过去的几十年里，世界经济的全球化已经达到了前所未有的水平。随着市场自由化的不断加剧，贸易壁垒被一一打破，企业正朝着国际化的方向稳步迈进，以便在新的市场不断增加市场份额，并且通过在具备丰富的生产要素和低廉的运作成本的地区设立分支机构，来获得成本优势。其结果是企业在全球范围内进行分散经营，深入到全球各地的价值链中，为全球范围内的各种客户提供服务。

背景与范围。 全球物流系统在商业活动的国际化过程中扮演着重要的角色。他们在全球价值链的成本效益和顾客要求的服务水平之间起着桥梁纽带的作用。所有这些无一不清楚地表明，物流是决定企业在全球竞争中成败与否的决定性因素。

此次调查的目的是从物流的角度去观察企业的国际化进程。建立国际化物流网络的挑战是什么？当进入某特定市场时物流经理所面对的问题是什么？他们该如何去应对这些挑战来确保企业全球化的成功实施？

本次调查在中德两国同时展开，对两国的企业全球化过程进行了问卷调查。为了使调查结果能够尽可能真实地反映出实际情况，我们对一些较为重要的国家和地区进行了多方面的分析，其中包括中国、印度、俄罗斯、南美、北美、东欧、西欧。尽管如此，本次调查并不仅仅只是单纯的市场调研。我们主要侧重于研究国际化进程中的行为模式以及进入国外市场时所制定的物流战略。

调查总结。 本次调查从宏观的角度对中德企业在国际化进程中的种种经历进行了资料搜集和经验总结，包括物流组织结构的设立，人员的雇佣，物流战略制定等各种活动。从中我们清楚地意识到，虽然国际化已经不再是新鲜事物，在国外建厂的总体环境也比以前好了很多，但是这仍然是一项充满了巨大的挑战的任务。

本次问卷的出发点是希望了解驱动企业开拓海外市场的目标是什么，这也是以后进行更深入研究的基础所在。调查结果表明，企业有着一系列的复杂目标，可以归纳为两大类：取得成本优势以及在国外不同各市场间均衡发展。因此在对物流系统进行设计时，既要考虑到适应本地的需求，同时建立的新设施又能紧密地融入到全球物流网络中去。

我们可以看到，一方面，先进的物流系统能够促进国际化进程，另一方面，国际化对时间、质量和成本的高要求也使得物流行业面临着挑战。因此国际化物流与各国内部物流有着很大的区别。当企业进入一个新的市场时，面临的最大挑战是基础设施的建设情况（这一点在发展中国家尤为突出）、安全问题、以及两国间的文化差别。能否实施良好的跨文化管理对于物流经理而言，显得越来越重要。从整合国内物流到建立全球化物流体系的过程中，总会出现各种各样的问题。关键是企业必须认识到不同国家的环境相差很大并且变化频繁。因此他们所采用的物流策略必须能够很好地应对这些复杂性和不确定性。需要明确的是，在国际化贸易往来中，主要的问题已经不仅仅是关税带来的

直接成本，而且还有由于运输过程中的种种不确定性所造成的额外库存成本。

本次调查的核心是对国际化进程进行分析。我们建立了一个简单的序列过程模型，用以表示从最初的战略制定，到在新市场中设立机构，再到新物流系统的最后实施以及与全球网络的集成这一连续的过程。中国企业完成这一过程需要14个月，而德国企业稍长约为18个月，这表明企业进入国外市场时所面临的严峻时间压力。

重视物流的作用，并让物流经理参与到企业决策活动中来，尤其是在进入国外市场初期制定战略决策的阶段，对于促进企业的发展大有好处。成功的企业大多都是一开始就对各种物流资源进行整合，这样不仅能够节约成本，还能更好地满足客户的需求，从而获得竞争优势。

在一个新的市场建立物流系统时，中国企业和德国企业的目标并不相同。中国企业关注于短期和具有成本效益的建设阶段，德国则更加看重最终系统的情况，来确保系统的可靠运行，时间的节省以及成本的减少。因此，德国企业更青睐于长期的计划和实施阶段。

在进行海外业务拓展时，应考虑三个物流方面的关键元素：

首先，物流经理必须确保物流系统具有高水平的柔性和敏捷性，以满足未能预知的未来需求。我们所访问的公司中大多已经具备了基本的柔性，但是其重要性仍然排在成本和时间之后。一个好的物流系统应当能够使企业在供应链的生产、采购、销售等各方面根据实际情况进行快速

调整。令人意外的是，尽管市场开拓失败的例子日益增多，但是很少企业有在他们的物流计划中包括因失败而退出市场的战略。

第二，为了应付国际物流系统的多样性和复杂性，物流工作者需要在国际标准流程与当地实际适应能力的冲突中进行平衡。全球物流标准化的好处已经广为认同，但还没有起到完全的标杆作用。企业仍然要在很大程度上使他们的物流系统能够适应当地的物流需求。

第三个因素是与物流服务提供商的战略合作。一直以来，中国企业的外包业务就比西方企业少。但是在开拓国外市场的过程中，如果与物流服务提供商加强合作，将加快物流系统的建立。因此，国际化将需要更高的外包程度和更大的外包活动范围。

为了降低全球供应链中存在的地理跨度和较大的文化差异带来的不良影响，需要找到一种整体物流管理策略，来对方方面面进行综合考虑，包括不同区域的生活方式、工作方法、技术设备、基础设施以及专业的国际贸易物流服务。在国外建厂的最重要的成功因素是认识到其多样性和复杂性，并确保工作人员经过充分的培训并且能够有效地完成工作。

研究目的和研究方法

本报告是研究"物流系统国际化"的成果，由德国柏林工业大学（TUB）物流系国际物流网络能力中心与华中科技大学（HUST）管理学院供应链与物流研究所共同完成。

研究目的。一旦公司的业务国际化，他们的物流系统就要相应地扩展。例如，在低成本国家建立新工厂，或是为了渗透新的区域市场而在当地开设销售公司或商店。这些会为物流管理者带来挑战，诸如设计、实施和监控正确的物流网络，以处理相应的货物流和信息流。

物流管理者该如何应付这些挑战，他们又会面临哪些来自组织内部和外部的问题？本调查着眼于企业在海外市场开展商业活动的全过程，包括从制定进入一个特定区域的最初战略决策，到当地物流系统的最终实施，直至最后整合成一个全球的网络。因此，我们分析了物流管理者运用的战略方法，通过比较中国和德国的管理者来分析他们使用方法的异同点。鉴于如今物流对企业效力的影响非常大，本调查试图找出是否存在某些特殊因素能导致企业的成功。

本研究仅限于不以物流为其核心竞争力的企业，比如制造企业和零售企业。除了物流服务提供商外，本研究涵盖了几乎所有不同的商业模式和不同的国际化目标。

研究方法。本研究项目结合了定性和定量的方法，全面回顾了物流系统国际化方面的文献和关于此领域最新发展的实际研究，并以此为基础形成本研究的总体设计。为了广泛而有效地了解国际化进程的模式，以及企业采用的战略和方法，研究采用了问卷调查的方式。问卷针对的是中国和德国的制造企业和零售企业的物流主管，被调查企业已有国际业务或是有意即将实行国际化。研究仅调查这些企业进入有限的几个目标地区和国家的全球扩张业务，这些是我们认为最重要的地区和国家，包括：北美、南美、西欧、东欧，印度，俄罗斯和中国。<所有区域参见图1>（选择这些为目标地区的细节参见第一章）

图1 – 调查方法的异同点

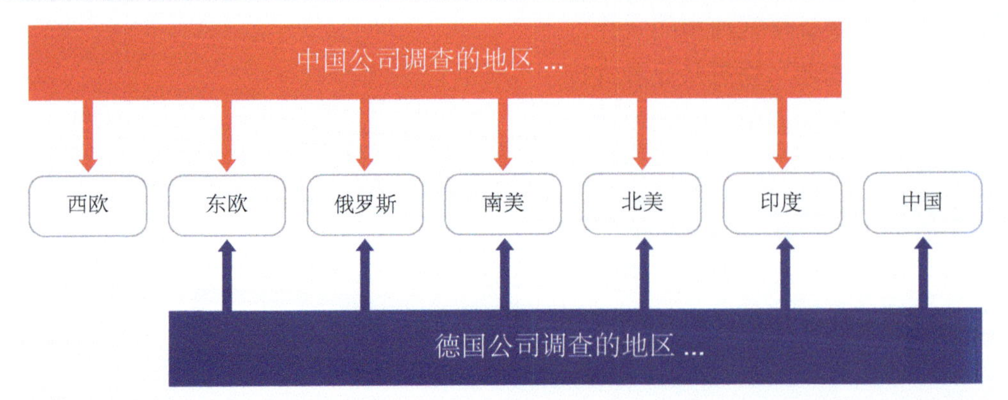

图2 - 样本图

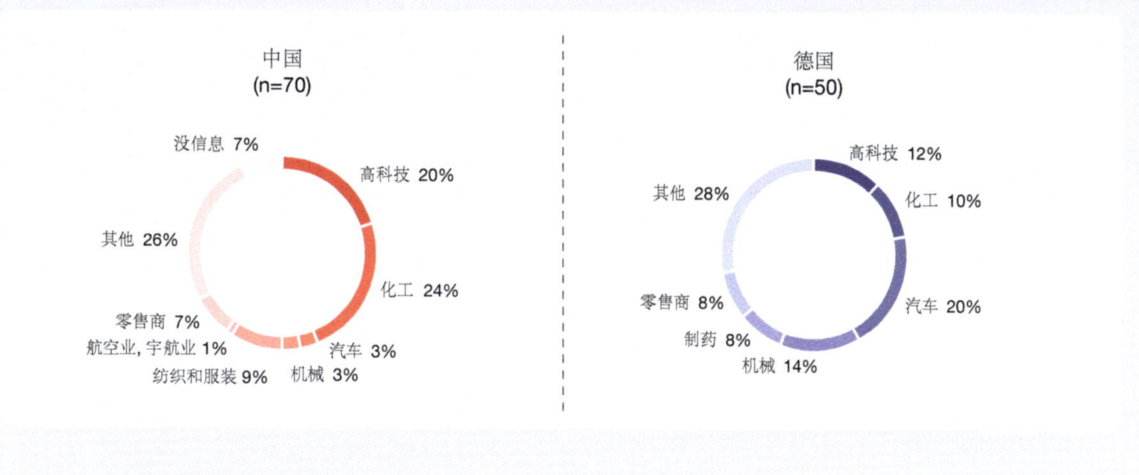

问卷中的观点来源于阅读文献时产生的想法和理念，在与从德国和中国挑选出的从业者进行深入的采访后，我们对这些观点进行了整理和检验。

接着问卷被翻译为德语，英语和中文三个版本，分发给中国和德国的企业，企业可以选择填写书面问卷或网上问卷。问卷调查于2006年6月开始，于2007年1月截止回收。向德国企业寄出1000份问卷，回收50份有效问卷。向中国企业寄出900份问卷，回收70份有效问卷。 其中20 % 是在线填写问卷. <参见图2> 图表显示了样本的行业分布.

对中德数据结果的分析由两个项目团队共同完成，采用实证研究通用的统计方法。与此同时，我们还与物流从业者进行讨论，增强了对数据的理解。

除了对结果的分析之外，本报告还提供来自多种渠道的、对选定目标区域以及特殊国际化课题的额外信息。

调查报告

插页目录

目录

图示目录

表格目录

缩写目录

3PL............ Third Party Logistics
APS............ Advanced Planning Systems
Benelux........ Belgium, Netherlands, Luxembourg
BRIC........... Brazil, Russia, India, China
CAGR........... Compound Annual Growth Rate
EDI............ Electronic Data Interchange
ERP............ Enterprise Resource Planning
EU............. European Union
FBU............ Fully Built Up
FCL............ Full Container Load
FDI............ Foreign Direct Investment
GDP............ Gross Domestic Product
ICD............ Inland Container Depots
ID............. Identification
IT............. Information Technology
KPI............ Key Performance Indicator
LSP............ Logistics Service Provider
M&A............ Mergers and Acquisitions
MERCOSUR....... Mercado Común del Sur (English: Southern Common Market)
NAFTA.......... North American Free Trade Agreement
R&D............ Research & Development
RFID........... Radio-frequency Identification
RMB............ Renminbi: official currency of the People's Republic of China
SARS........... Severe Acute Respiratory Syndrome
SCM............ Supply Chain Management
SMEs........... Small and Medium Sized Enterprises
SUV............ Sport Utility Vehicle
TEU............ Twenty-foot Equivalent Unit
UNCTAD......... United Nations Conference on Trade and Development
WTO............ World Trade Organisation

物流系统的国际化
– 中国和德国公司怎么进入国外市场

1 序论 – 商业活动的国际化

在上个世纪，跨国经济活动的雇用和执行状况得到了显著提高，这也导致了商品流通出现了前所未有的全球化趋势，以及价值链遍布世界各地的蓬勃发展。

世界经济的一体化。如图所示，世界生产总值的发展同世界商品出口的发展的比较<参见图3和表1>说明了世界经济的联系正在不断增强。市场的自由化、运输成本的降低、以及信息和通信技术的日新月异不仅导致国际贸易的增长，即通常意义上产品的海外销售，而且还导致了生产系统的全球连通。在分散的商业活动日益朝着全球集成发展的当今，物流成为了主要的推动力量，通过设计、管理和监控全世界货物和信息的流动，物流被当之无愧地看作是全球化现象的重要标识。

全球市场的变化。尽管北美、西欧和日本几乎占据了世界生产总值的半壁江山，并且全球财富500< 参见图 6 >强中大部分的公司都来自这三个地区，但当前的经济数据预示着老牌的三巨头已经过时了。被称为"金砖国家"（巴西，俄罗斯，印度，中国）的新兴市场快速发展，给世人留下了印象深刻的高增长率。再考虑一下这些国家占世界人口的比例和占世界生产总值的份额之间显著的不均衡吧< 参见图 3 >。它们拥有在未来持久和持续增长的巨大潜力。除此之外，比起三巨头，像中国这样的经济体拥有更先进和更科学的竞争力，它们不再仅仅提供低成本的劳动力，而是表现出参与国际市场竞争的能力和雄心。它们改变了发达国家和发展中国家之间的传统分工，使得经济的全球化朝着双向发展：从发达国家流向发展中国家，同时也从发展中国家流向发达国家。

进入国外最重要的目标市场。根据地区经济同世界经济的关系，本调查确定了七个国际化进程最重要的目标地区和国家，分别是：北美、南美、西欧、东欧、俄罗斯、印度和中国。

考虑到中国、印度和俄罗斯现有的高度重要性和未来积极的发展前景，本调查将它们作为单独的国家进行分析。这三个国家显示了非常强大的经济潜力，尽管它们现在的GDP之和并不高，仅占世界总值的8%，但在过去的十年里，它们的增长率远远高于那些经济巨头<参见图 4>。比如，中国几乎连续30年保持两位数的增长趋势。印度的平均GDP增长在过去十年翻了一番。

为了使本调查保持适当的复杂度，将其他国家归入北美、南美、西欧和东欧这四个地区。尽管这些地区的发展水平（政治情况和基础设施等）存在一定的差异，但是每个地区一般都自视为在经济，政治（"欧盟"，"南方共同市场"）和文化上独立的一个区域，并且得到了全世界的公认。西欧和北美代表发达地区，具有低增长率和相对较高的人均收入，但是全球对外直接投资（FDI）进出量的份额较高，而且这些经济体在市场上拥有数目最多的跨国公司<参见图 6和图 7">。

南美和东欧的经济一样被认为是新兴的或发展中的市场。人均GDP水平仍然远远低于发达地区，且增长率也落后于中

图3 – 世界出口总值与世界生产总值的发展，国与地区所占份额的比较

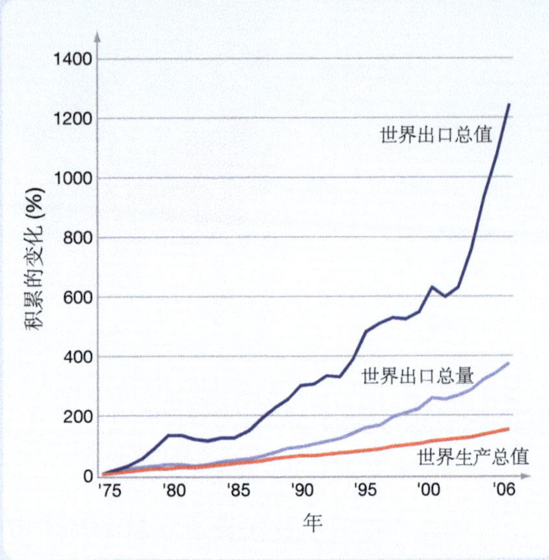

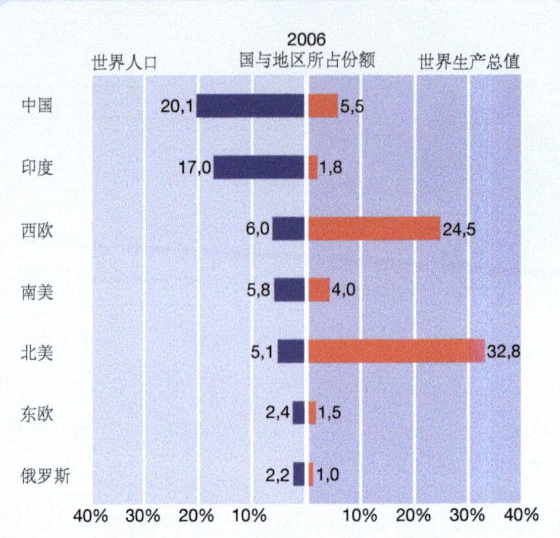

图4 – 世界生产总值的发展，生产总值年均复合增长率与财富的比较

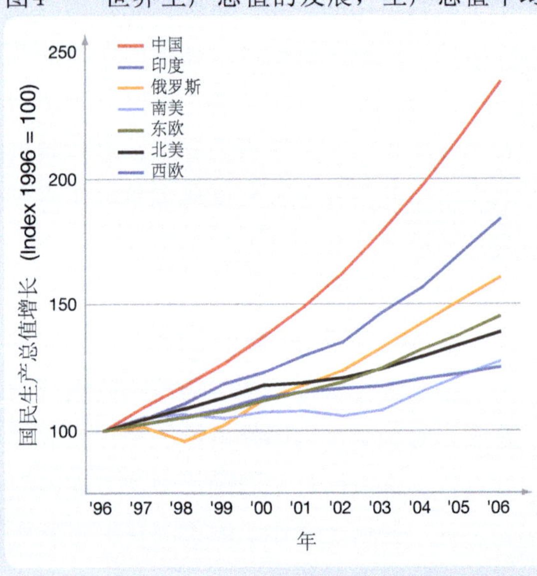

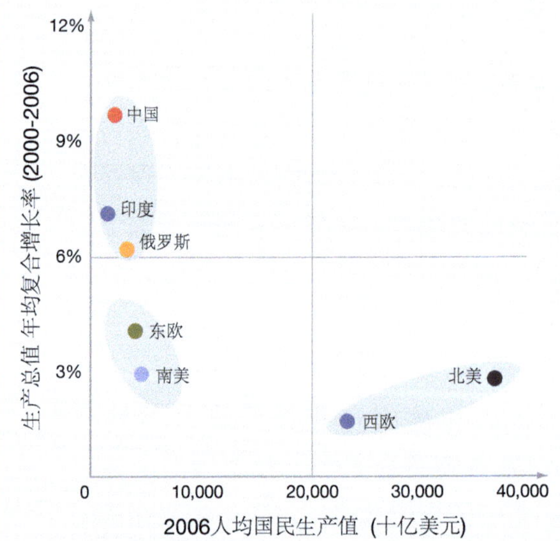

表1 – 特定地区和国家的重点数据

	国民生产总值 2006 (美元/人口)	生产总值年均复合增长率 2000-`06 (%)	直接投资 2005 (百万美元)	直接投资年均复合增长率 2000-`05 (%)	出口 2005 (十亿美元)	出口年均复合增长率 2000-`05 (%)	进口 2005 (十亿美元)	年均复合增长率 2000-`05 (%)
西欧	23,480.35	1.7	397,066.49	-12.64	4,549.04	8.5	4,408.96	8.5
东欧	3,549.05	4.4	50,651.20	6.68	478.86	18.7	505.21	19.3
北美	3,7387.89	2.8	132,264.90	-12.99	1,551.17	2.1	2,134.13	4.9
南美	4,008.56	2.9	44,697.36	-7.66	354.53	12.7	268.60	7.0
俄罗斯	2,608.64	6.1	14,599.61	5.53	268.32	18.6	164.62	21.4
印度	612.24	6.9	6,598.00	3.96	165.49	21.1	194.80	23.8
中国	1,579.07	9.8	109,073.69	4.78	836.89	24.5	712.09	23.2

图5 － 中国和德国的互相经济关系

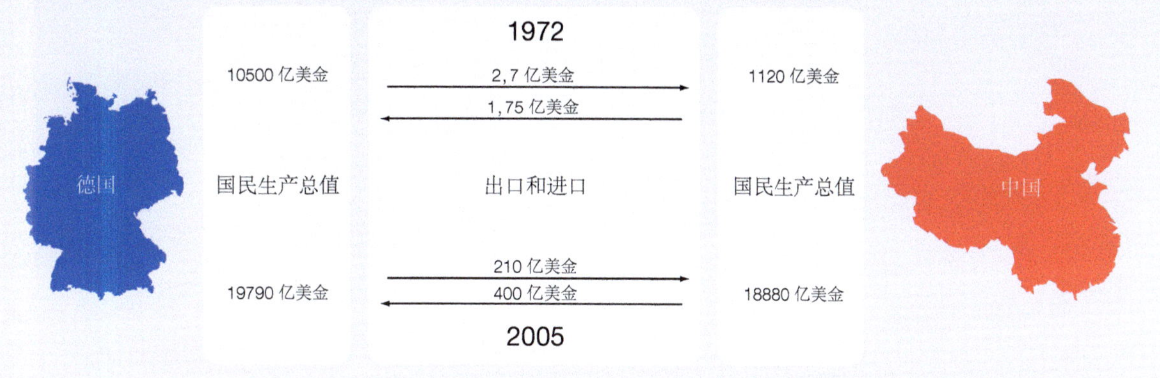

国、印度和俄罗斯这些国家。从它们在全球FDI流入量的份额中可以看出，它们在未来全球市场上将有着非常积极的发展前景。

以这些国家和地区作为探讨的主题，本调查为进入国外关键的目标市场提供了全面的分析，同时我们的数据表中保存着大量直接的数据。（想更多地知道关于个别地区和国家的情况，请参见相关的部分）。

中国和德国企业如何进入新的市场。本调查着眼于中国和德国的企业国际扩张中的物流战略。这两个国家在国际贸易中有极高的参与度，是世界经济的关键参与者，并且这两个国家在全球化中也是明显受益的<参见图5>。此外，中德之间相互的经济联系也很紧密：德国是中国在欧洲最大的出口市场，占中国在欧洲对外直接投资的最大份额；中国是德国在全世界第二大出口市场，占德国在亚洲对外直接投资中的主要份额。

德国的跨国企业代表了立足于发达市场并在国际化进程中积累了数十年经验的企业。而另一方面，中国企业代表的是立足于新兴市场的企业。这些企业在国内受益于低劳动成本的投入，在国内市场的竞争中已经获得了成功，现在同时向发达国家和发展中国家的新市场进军，这些新兴的中国企业具有导致全球化竞争发生重大变革的潜力。

本调查基于120个企业物流主管提供的、在全球化进程中制定物流战略细节的样本，对关于进入国外市场的多种相关课题进行了深入研究，明确了中德方式的异同之处，提供了全面了解不同目标市场所需的实际情况和数据。

图6 - Fortune Global 500 的分布 - 2006 的公司

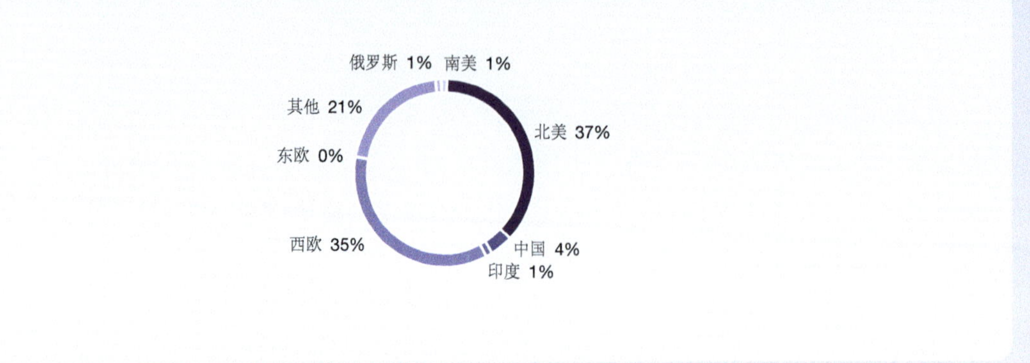

Source: 2006 Fortune Global 500

图7 - 各国家的直接投资 - 2005年（百万美元）

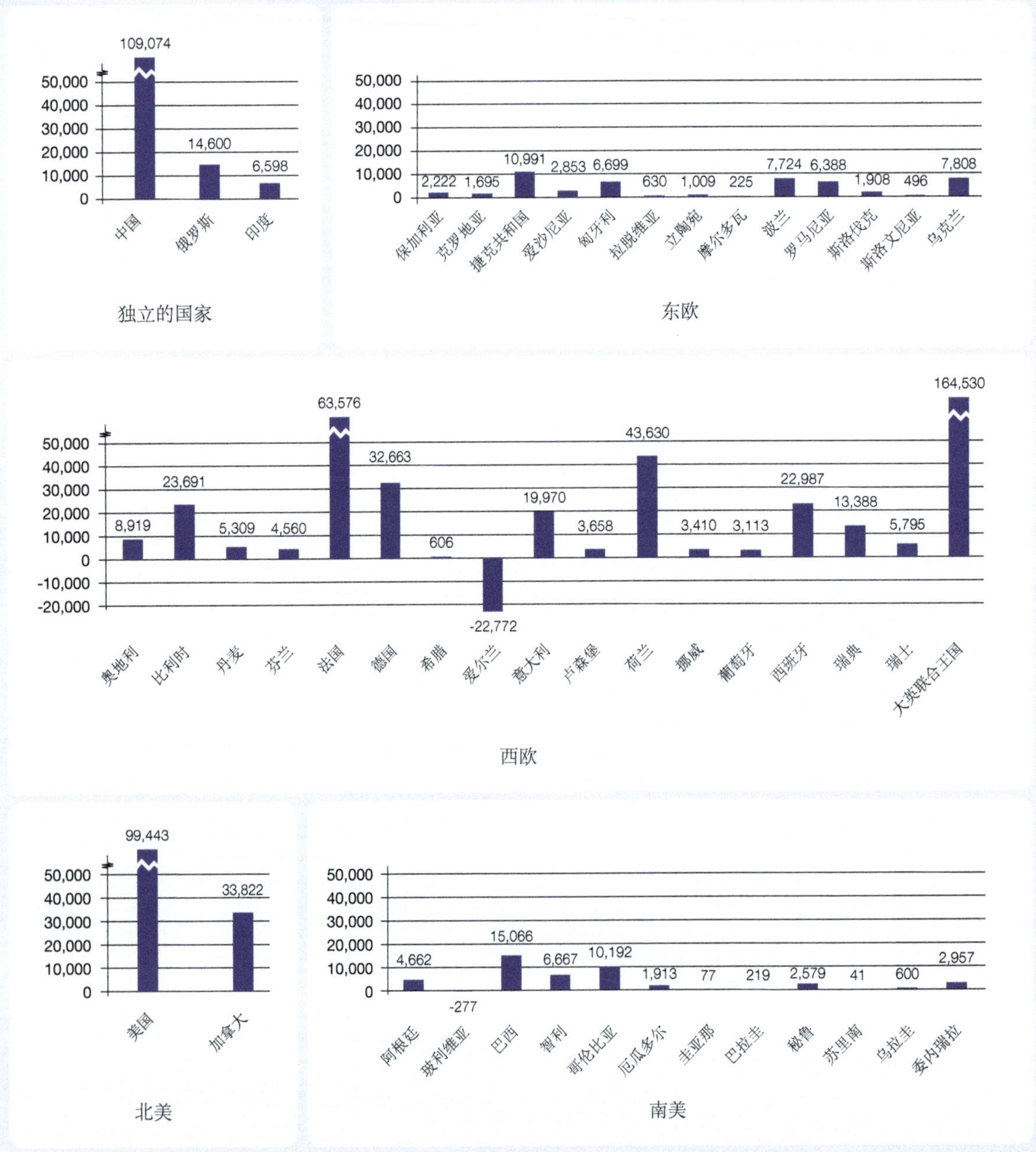

Source: UNCTAD databases

中国一览

　　自从1978年实行改革开放以来，中国经济实现了快速增长，目前已经超过英国成为世界第四经济大国。在过去的25年里，中国大陆经济每年平均实际增长率超过9.5%，成为经济大国中经济增长记录的保持者。2000年，中国GDP仅占世界GDP的4%，而预计到2025年，该比例将升至的11%。

　　从2001年至2005年间，中国贸易年均增长率近29%，而在2005年，中国更是成为仅次于美国和德国的世界第三贸易大国，并且有望在21世纪20年代成为世界最大贸易国。因为中国的出口贸易总额占到GDP的72%，所以中国将会更多地从贸易自由化进程中受益。从2002年到2005年的四年间，中国吸收了近2300亿美元的外资，这些资金的流入加速了中国与世界经济的接轨。

　　在经过了15年的谈判后，中国在2001年正式加入WTO，标志着中国在经济全球化进程的战略决策中迈出了重要的一步。国有企业的地位逐渐削弱，不具备竞争优势的企业就会被市场淘汰，与此同时，私有企业则因为中国加入了WTO而获益良多。首先，私有企业可以拥有和国企一样的市场机遇。其次，私有企业通过商业合作能很快掌握国外先进管理、技术以及资金。

　　中国在加入WTO后的六年中发生了很大变化。市场经济、自由贸易和投资的观念深入人心，此外WTO的核心理念如透明化、责任管理、国民待遇等也被民众所广为接受。中国的入世对世界经济面貌也产生了很大影响。中国已经成为世界制造中心和世界经济增长的动力，如今全球采购业已聚焦中国。

　　中国拥有七亿四千五百万劳动力，主要在分布在三个领域：农业（50%）、工业（22%）、服务业（28%）。然而，中国的失业率却很高，在城市达到10%，在农村则还要更高一些。温家宝总理曾经说过，城乡差距越来越大，农村经济停滞、教育落后、医疗卫生条件跟不上和高失业率的问题已经越来越尖锐。

　　因此不难理解为什么全球消费品公司一开始就致力于在中国三大城市发展：北京、上海和广州。这三个中心城市拥有2900万人口，包括中国最丰富和成熟的市场，拥有相当于13%的国家可支配收入。然而，几乎没有什么国际企业关注分布在中国各处的约12000个小城镇，这些城镇的家庭总收入已经比那些所谓一级二级城市高出50%。预计在将来的20年中，家庭年收入超过35,000元的小城镇家庭数量的年增长率为7%，约为760万个，而一二级城市的预计年平均增长率为5%，约为660万个。总共约有4亿人（约占全国总人口的1/3）生活在这些小城镇，这些小城镇正是城乡之间的过渡桥梁。

中国作为世界第三大贸易国，其物流业已经变得越来越重要。中国政府在第十和第十一个五年计划中都强调，要发展物流业以促进和支持经济的可持续发展。目前，其运输成本占了整个物流成本的近55%。因此，中国政府已经将注意力转移到削减运输相关的开支上来。为了改善运输成本过高的状况，中国政府在2005年投资了7750亿人民币用于基础设施的建设，以解决交通运输网络中的瓶颈问题。这一数字比前一年增长了23%。

为了改善内陆交通，相当多的投资被用于高速公路的建设。在2001年至2005年间，中国的高速公路建设投资以每年6.7%的速度增长。此外，中国正致力于扩大其港口吞吐能力，以满足日益增长的需求。以标准集装箱吞吐量计算，中国的深圳和上海港已位居世界四大集装箱港口之列，仅次于新加坡和香港。

航空运输业在中国亦高速发展。在过去二十多年里，中国十大机场中有七个机场都保持着两位数增长。2005年中国的航空货运业增长了11%。据估计，到2010年将增长50%以上。事实上，中国的交通运输网络是世界上最大的交通运输网络之一。即使在欠发达的地区，如中国西南地区，到2010年，机场的数量也将是现在的两倍，即由24个增加到48个。

当然中国的发展也有不足的一面。中国经济的发展依赖产品出口，这就使中国经济容易受到外部经济波动的干扰。物流服务市场更是相当分散，并且发展滞后。有关专家估计中国现有物流企业的总数约为7万个，但是没有任何一家企业的市场占有率能够超过2%。GDP中约有20%用于物流费用支出，这个费用是美国的两倍，产品成本中的40%都是与物流有关的费用。以中国对美国的出口为例，出口量明显高于进口量。这就使得贸易中运输费用的价格结构会受到空返中国大陆的集装箱的较大影响。

有些人担心中国真正面临的挑战是更深层次的市场开放和中国领导层的能力，以及其经济应对挑战的能力。另外有些人则期望中国能提供更多的市场机遇，并且在诸如知识产权保护等问题上有所进步。

2 国际化的战略目标

物流战略是公司战略的子系统，是实现公司目标的保证。因此，在调查的第一步，企业在国际化进程中的战略目标通常是进行深入分析的基础。由于在不同的地区，公司的目标可能存在差异。因此，目标决定了当公司进入一个特定目标地区或国家时希望得到的是什么。

并不是所有的目标都可以被清楚地区分出来，因为这些目标可能是紧密联系的，并且有复杂的相关性。一般而言，企业在国外开设工厂的目标都可以归纳到一个目标集中，见图 8。图表显示了每一个目标对于中德企业以及其他特定地区或国家的重要性。

进入国外市场的目标。通常而言，进入国外市场的两个主要动机是挖掘发展的潜力和降低成本的潜力。

然而，有趣的是发展方面的价值比降低成本更重要。当企业利用全球化带来的机遇时，全球化便巩固了企业的市场地位。公共媒体上常见的反对言论声称，全球化成本的压力使企业将生产转移到国外，从而使得发达国家的失业率上升，但这种压力在当今企业的国际化进程中只起到非常小的作用。国际化首先意味着影响全球化扩张的机会增多，这才是企业获得的最大优势。

过去，那些明确以成本为中心的企业常常在低成本国家的"经济特区"或"自由贸易区"设厂，以享受较低的制造成本。生产的商品则主要用于出口。通常这些经济特区直接与国际港口或机场相连，

并且提供国际货物的加工服务，因此促进了物流的发展，也使得接下来的网络构建变得相对简单了。现在，由于在许多情况中，目标集的设定变得更加复杂，而且许多企业同时遵循成本降低和发展的双重目标，新的海外机构必须同时满足所在的新市场以及向其他地区出口两个方面的需求。所以，迫切需要复杂的物流系统，将区域的供应和分销部分集成到全球的物流网络中去。

中德目标集的比较 。中德企业国际化进程中目标的类型存在一个明显的差异：德国企业的目标几乎都集中于发展和成本，而中国企业对这些目标回答的分布却比较平均。(参见图 8 中的两条曲线)。

如第一章所述，像德国这样的发达国家发展缓慢或停滞，而发展中国家的购买力却在激增。同时，我们必须注意到：发展中国家还拥有巨大的成本优势，使得它们能够参与到那些有活力的市场的发展中去。这对于德国这样的国家具有非常重要的意义。

另一方面，中国企业有更广范的目标。国内竞争的日益加剧，国外市场机会日益增多，进入国外市场可以获得更多的技术创新、专业知识和专业人才，也可以通过地区和市场的分散化经营来平衡风险，从而使企业获得进一步的机会。

尽管如此，发展对于中国企业和德国企业而言都是最重要的，尽管仍存在着明显的差别：中国企业寻求利用国内的成本优势和国外市场相对高的价格来获取更高

图8 – 国际化进程的主要目标

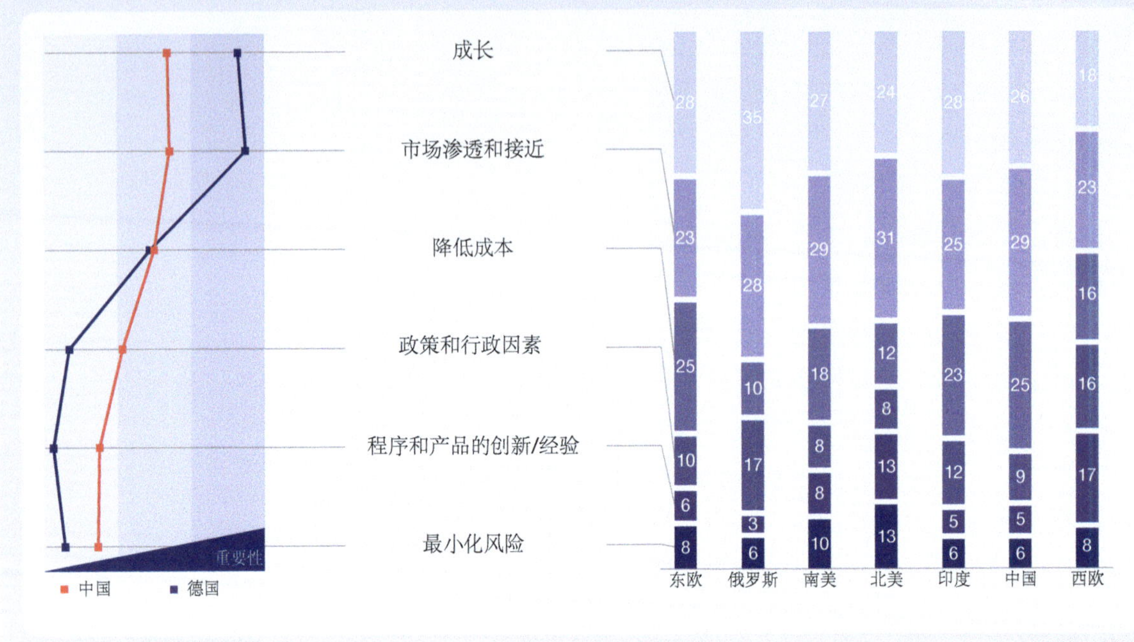

的边际收益；德国企业则由于国内市场相对饱和，希望扩大客户群以获取规模经济效益。由于近年来产品平均研发成本的日益增长，产品生命周期逐渐缩短，广阔的客户群能保证企业更早达到盈亏平衡，获得更高的回报。因此，它们甚至认为发展中国家市场的低价格水平导致了较小的边际收益。

尽管中国总成本的情况更有利些，但中国企业却同他们的德国同行一样重视成本降低，乍看之下，这一点似乎有些奇怪。但是，在发达国家制造某些特殊商品可能比在中国制造更节省成本。尤其对于有高质量要求的知识密集型产品而言，发达国家具备的生产要素显示出明显的优势。受过良好教育和培训且精通操作流程的工人，能够减少浪费和返工，提供较高的可靠性，从而降低了质量成本，抵消了相对较高的劳动成本。因此对于中德企业，进入国际市场同样有利于全面改善成本结构。

国际化商业活动的另一个目标是降低风险，这可以通过在不同的地区进行分散制造来实现，从而保证可靠的运作，避免币值波动。此外，在国外自营制造还可以防止知识产权受侵犯，而外包给低成本国家制造时却很容易发生这种情况。

区域间的差异。低成本和购买力日益增长所带来的巨大消费市场，给中国和印度吸引外资企业来此设厂带来了双重优势。因此，企业设立的机构要同时满足当地和出口的需求。

东欧国家的情况也与之类似。尽管这些国家的增长率相对于中国和印度低一些，但是经济前景依然是很乐观的，越来越多的东欧国家加入欧盟也使得其未来具有很高的期望值。大多数东欧国家的劳动力成本依然很低，但是某些地区和行业领域的工资也赶上了西欧的水平。自由的税收制度和生产成本为进入市场创造了有利的总体环境。东欧国家一个明显的优势是

在地理上接近西边大陆的富裕国家（欧盟15国，瑞士和挪威），这使得他们对于那些邻国很有吸引力。

目前，在南美设厂的主要动机是为了满足国内需求，以及参与到像阿根廷或巴西这样的积极市场的发展中去。进行低成本制造用于出口并不是最重要的目的。

俄罗斯吸引国外制造商和零售商进入的原因在于消费者日益增长的购买力。与东亚和印度这样的新兴市场不同的是，俄罗斯并不是世界市场中制造成本较低的国家。

西欧和北美最有趣，这是因为它们的消费者具有极高的购买力。正如前文提到的，只有高技巧性的特殊商品才能节省成本，相对而言，流程、产品创新和技术对在这里设厂更加重要。

在大多数情况下，是否在一个特定国家设立海外机构，最终还取决于政策和行政因素。这包括补贴、减税、关税和非关税贸易壁垒。比如当地的规章和管理程序，会妨碍一个公司向特定市场的出口。这些措施对进入国外市场的获利能力有着重大影响，政策和行政因素在俄罗斯、印度、中国、东欧和西欧尤其具有决定性作用。

3 物流和国际化之间的关系

物流和国际化彼此紧密相连。一方面，作为国际化的关键驱动因素之一，物流促进了国际化的发展；另一方面，国际化正在以空前的速度向旧的物流规则发出挑战。

很明显，如果没有全球运输和贸易系统的发展，当今的全球化不可能达到如此高的程度。如上图所示，在上个世纪中，运输费用和贸易费用已有了明显的减少。最重要的两种运输模式，海运和空运的费用，在过去的70年里已经分别降低了65%和88%。与此同时，商品的运输速度和信息的传递速度也达到了无法预料的水平，就像今天全球网络能够提供的能力一样。

这些明显地促使公司在靠近海岸的地方建立工厂，并把海外业务整合到高度复杂的价值链中。同时，如何设计和管理这些复杂的货物流和信息流，成为了全世界的物流管理者亟需解决的问题。

现有的全球贸易和先进的运输系统为全球业务提供了一个重要的先决条件，但这并不意味着进行全球业务是一件简单的事情。为了成功地从事国际商业活动，需要掌握一整套复杂的管理方法，包括物流原理和方法论。

3.1 物流对企业成功的重要性

为了在国际化背景下理解企业物流的重要性，首先必须从宏观角度来看看物流在当今的商业中扮演的角色。

在过去的几十年里，物流规则不断地发生着变化，如图9所示。

物流在过去的发展。在上世纪六七十年代，传统的物流指的是原材料以及与商品相关的任务和职责，包括运输、装卸、仓储，包装以及配送等。在这种情况下，原材料和零件的可获性就十分重要。那时，人们并不认为物流是影响企业成功的主要因素，它本身不是一门学科，而是以孤立的方式体现在企业的组织结构中，与其它部门没有适当的联系。在八十年代，人们对物流的认识从一项职能转变为以流动为导向的手段，此时物流的目的是通过采购、生产以及销售等环节对以往分散的业务进行整合和协调，以达到对流程的优化。

在九十年代，职能集成的概念得到了发展，货物和信息的流动也越来越一体化。通过这种一体化的方法，人们能够对整个流程链进行跨职能的整合优化。IT行业的飞速发展也使得在流程链之间以及内部信息不足的现象得以缓解。

3. 物流和国际化之间的关系

图9 - 物流的发展

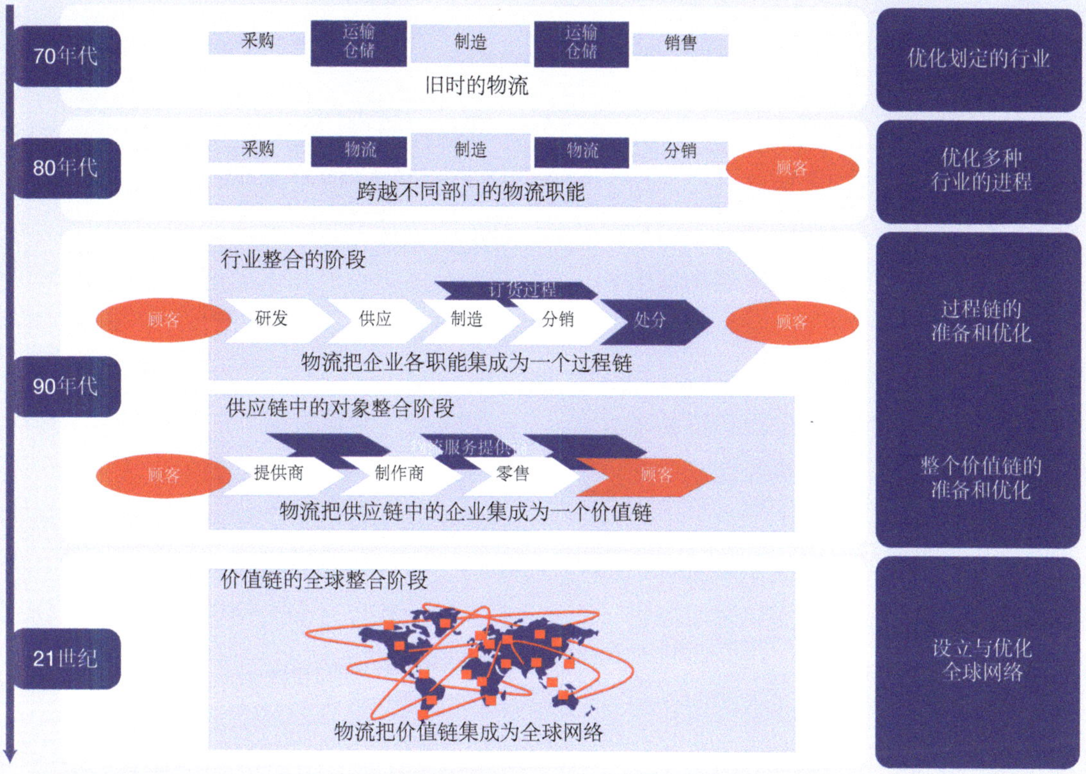

Source: Baumgarten et al. (2004)

在战略业务层面上，物流已经成为了一项关键的职能。事实证明，物流能够迅速且灵活地满足企业快速增长的需求，从而使企业获得巨大的竞争优势，特别是在服务驱动和基于时间的竞争中，这种优势更加明显，已经在越来越多的市场中体现出来。

进入新千年后，公司开始在企业层面进行跨职能和整体的协调。货物和信息的流动不仅在企业内部各职能部门之间，同时也在各个企业之间得到了控制和协调，目的是为了使供应商和顾客建立更紧密的联系。这种结合由于供应链管理这一概念变得更加先进了，供应链管理的基础是广泛普及的企业资源计划(ERP)系统，并通过电子数据交换(EDI）进行连接。日后的高级排产系统（APS）将进一步促进这种发展。因此，这种多个企业的整合将不仅仅局限于国内——也就是说，现阶段物流革命的特征是对整个全球网络进行组织和优化。

企业物流的现状。现在，很少有企业能够在全球范围内对网络进行整合和优化。在大多数发达国家，企业通常采用以过程为导向的物流集成方法，主要与他们的直接供应商以及顾客有信息交换，并且对供应链上下游的某些关键点进行监控。但是整个全球网络的整合还处于起步阶段。与其他迅速发展的国家相似，一般中国企业的情况是不同的。它们仍然把物流看作一项独立的功能，主要由企业的各个职能部门单独地从事运输和仓储工作，类似于七八十年代的西方经济时的方法。

那些已经进行或即将进行国际化经营的中国企业，则对物流有着更为先进的认识。从这次参与调查的中国企业中，我们得出了一个与预期并不相同的结论：那些

即将参与全球竞争的中国企业和它们的西方伙伴一样，都认为物流很重要，大多数成功的中国企业在进行国际化时，都会在物流方面进行系统的规划。事实上，这些企业在进入市场时走的是"本土化"的路线，使他们可以很好的适应当地情况，这就使得这些企业有机会从战略层面上对物流进行系统的规划。所以，值得一提的是，不要因为看到一般的中国企业都着眼于国内的竞争，就低估了那些即将参与全球角逐的中国企业。

物流在当今企业成功中所扮演的角色。参与调查的90%的企业都认为物流对于竞争至关重要，因为顾客要求企业能够在物流性能和可靠性方面提供高水平的服务。物流服务水平的高低将直接影响顾客的最终购买，这是因为产品与可获性、提前期以及其它服务等方面紧密相连。调查显示，中国企业的市场竞争大多在商品市场领域，价格是最重要的销售标准，因此它们更关注成本（中国：92%，德国62%），而德国的企业则更注重其它方面，例如：技术熟练程度、质量以及服务，因此相对于中国企业，参与调查的德国企业在竞争时更加注重服务水平（中国：65%，德国：79%）。

至关重要的一点是，超过70%的管理者指出：激烈的国际竞争导致顾客希望获得高水平的物流服务，然而他们却不愿意为服务额外付费。

因此，随着顾客的期望越来越高，企业在成本上的压力也越来越大，物流系统的有效性和高效性对企业的竞争也显得越来越重要。

印度一览

印度共和国北靠喜马拉雅山，南临阿拉伯海和孟加拉海湾，约有11.1亿人口，占世界总人口的17%。与中国相邻的印度被认为是亚洲第二新兴国家，并且在国际商业中的地位变得日益重要。其对外贸易总额已经占到GDP的三分之一，2006年全国进出口贸易总额（4月-12月）增加了36%，分别达到1312亿美元（进口）和895亿美元（出口）。

在过去的几年里，印度的经济一直保持着8%的年均增长率，但尽管印度经历了15年的经济增长，她目前仍然是一个发展中国家，约有占印度总人口（11.1亿）四分之一的人仍然生活在贫困线以下。印度经济在近几年经历了一个很大的转变，从原来的初级产品生产过渡到二级、三级产品生产。2006/2007年度（2006-04-01至2007-03-31）的GDP达到9230亿美元，其中大部分是来自服务业和工业企业。2006年服务业总收入占到了GDP的55%，其中包括零售、宾馆、运输、保险、金融、公共和个人服务等。尽管印度是仅次于中国的世界第二粮食生产国，但是其农业收入仅占GDP的20%左右，占世界粮食市场份额的1.5%，她在世界范围内还处于从属地位。虽然印度农业对于在世界范围内来看无足轻重，但是农业却是一个国家人民的粮食支柱。从2005年到2006年间，印度再加工和次加工增长率为10%，而初级生产增长率却由原来的6%降到3%。

印度向来以IT技术闻名，被认为是离岸外包服务（如客户支持）的理想选择。除了服务与IT技术之外，印度在制药、航天技术、生物科技等方面也颇具竞争力。就在不久前，许多跨国公司如汽车制造企业（福特，现代，本田，宝马）纷纷在印度建厂。　印度政府为了减少对离岸外包业的依赖，正在扩大他们自己的生产线。

印度因为其社会文化的构成而成为最有潜力的采购市场，而促成这些的是国民熟练的英语、稳定的政府结构、大量资深工程师和技术工人。印度政府鼓励外商投资他们的生产和基础设施，并且制定法律允许外国公司全资注入他们某些特定工业部门的辅助部门。

印度的经济活动主要集中在以几个大城市为中心的商业圈，如新德里，孟买，班加罗尔，钦奈，加尔各答，艾哈迈达巴德，海德拉巴。在过去几年中，单是印度首都新德里接受的外国直接投资就超过了外资投资总额的三分之二。西海岸的孟买地区是全国的金融中心，这个地区贡献了将近40%的GDP，而在南部IT新型城市班加罗尔，其软件出口总额约占全国软件出口总额的90%，同时它也是航天和生物技术企业的聚集地。在南部的孟加拉海湾，马德拉斯和钦奈在城市周围建立了大型汽车工业生产基地。这里是许多跨国汽车生产企业如福特、雷诺、奔驰、现代、本田和宝马在印度的辅助生产地，也是他们的转包商所在地。离钦奈不远，生物工业在海德拉巴建立了所谓的"基因谷"。此外药品制造业（主要是通用药品）也在西北部的艾哈迈德巴德建立了基地。

正是由于这些地区的工业不断发展和壮大，所以需要高标准的物流服务来提供高效的、复杂的供应链。跨国物流提供商投入了大量的资金提高仓库物流能力。目前当地的物流提供商还不能在提供全国性的服务上与西方国家抗衡。

这些地方的集中发展也从几个方面给印度带来了挑战，发达的经济和高收入只是使少数人受益，农村条件并没有因此而改善，所以导致大量移民涌入到这些发达城市，造成城市周围贫民区的迅猛增加，而收入分配的不平等将会导致社会动荡。

世界银行的调查显示，投资的主要障碍是政府阻力、腐败和基础设施的匮乏。

尽管印度的的基础设施处于极其糟糕的状态，但截至目前为止，与最具竞争力的邻国中国相比，印度政府在基础设施方面的投资仅为中国在基建方面投资的很小一部分。中国每年对基础设施建设的投资接近2000亿美元，而印度只有280亿美元。

目前，印度的大多数铁路线路的运营仍然停留在1947年刚从英国殖民地解放时的技术水平，然而这样的铁路却是印度货运和客运的主要方式。鉴于政府的交通运输相关政策是补贴客运而加重货运的负担，因此现在货运方式已经明显转向公路运输。

公路是印度主要的运输方式，尽管公路密度与美国相当，但是路况却相差很多。只有不到10%的公路能够实现单向双道行驶，这其中仅有1%的公路可以称其为高速公路。40%的印度农村在天气不好时不能实现通车。由于长时间的堵塞和糟糕的路况，卡车的平均车速仅为30-40公里/小时。更糟糕的是民众缺乏遵守交通规则的意识以及公路上存在各种不同的群体。大多数公路包括高速公路在内，卡车司机必须面对各种群体如家畜，马队，孩子和骑自行车的人。为了改进全国公路系统，政府最近正在建设"金四角"，一旦建成，将连接德里、孟买、钦奈和卡洛塔四座城市。

印度有12个主要港口，为了适应国家日益重要的对外贸易活动，这些港口吞吐量必须大幅度提高（2倍）。目前，印度的集装箱平均搬运时间是世界平均时间的3-4倍。 现在印度还没有与国际贸易路线接轨，大批量的货物必须先运输至新加坡然后再转运至目的地。

设施的落后，国内税收（物流外包服务税收为12%）和关税使第三方物流提供商很难为客户提供高水平的服务。尽管如此，印度物流市场的年增长率仍然有望达到18%。

3.2 物流和成功的国际化活动

商业活动的国际化意味着商业活动将会更加无局限，以及整个价值链中的企业在地域上越来越分散。与那些集中在当地的运作相比，国际化的商业活动导致物流在全球网络中的角色更加重要，这是因为物流是用来处理由商业活动产生的货物流和信息流。全球供应链依赖于物流，物流费用在总费用中也占有很大的比例。与此同时，顾客对物流服务和性能的要求也越来越高〈参见3.1章〉。由此，我们可以得出一个结论：物流是构成任何国际化活动的重要支柱。

同时，需要确定的是公司和管理者对物流国际化的重要性应该达到何种认识程度，以及对于进入国外市场相关的项目是否考虑充分。

物流在国际化中扮演的角色。如图10所示，超过80%的受访企业认为物流是国际化成功的重要因素，物流需要与公司的其它职能密切互动。三分之一的德国企业认为物流是企业进行国际化最重要的推动力量，超过三分之二的中国管理者持相同观点。

样本中大约15%的公司遇到过这样一种状况：由于地区的物流条件较差而不采取进入该市场，例如：一家公司决定不进入东南亚的某个地区，是因为该地区的运输和仓储在控制温度方面存在不足。

在德国，不足20%的企业认为物流对于企业进入一个新市场时不重要，而在中国，这个比例则更低。

这显示了在全球化的背景下，相对于德国的伙伴，中国企业更加注重物流。可以归纳为以下几种原因：首先，中国企业在刚开始参与国际化竞争时遇到了更多的挑战。由于中国的基础设施还不够完善，物流服务市场还不够发达，出口时的装货容量经常短缺。从开始至今，物流功能在中国企业中扮演着重要的角色，特别是那些参与全球化的企业。与中国相比，德国的出口活动是在一个更有利的环境下进行的，因此，那些即将参与全球竞争的中国企业比德国的企业更注重物流。

其次，中国企业更多地集中于低端产品的生产，这就使他们面临着更大的成本压力。由于德国的企业能够承担更高的物流成本，因此，相对于德国企业来说，中国企业物流成本的高低就显得更为重要了。

物流在不同地区的重要性。除了发现受访者普遍认识到物流对国际化的重要性，我们还发现他们意识到物流在不同地区的重要性存在差异。尤其是在中国和印度，他们认为物流运作是一个非常重要的成功因素。

德国企业认为，作为新兴经济体的典型代表，中国和印度的物流决定了企业的成功与否。对中国企业来说，在像北美以及西欧这样发达的市场里，只有高效的物流组织，才能满足这些市场里的顾客对物流服务的相对较高的期望值。

总而言之，物流管理者们都认为优质的物流是国际化成功的基本条件之一，并

图10 － 国际化进程中物流的重要性

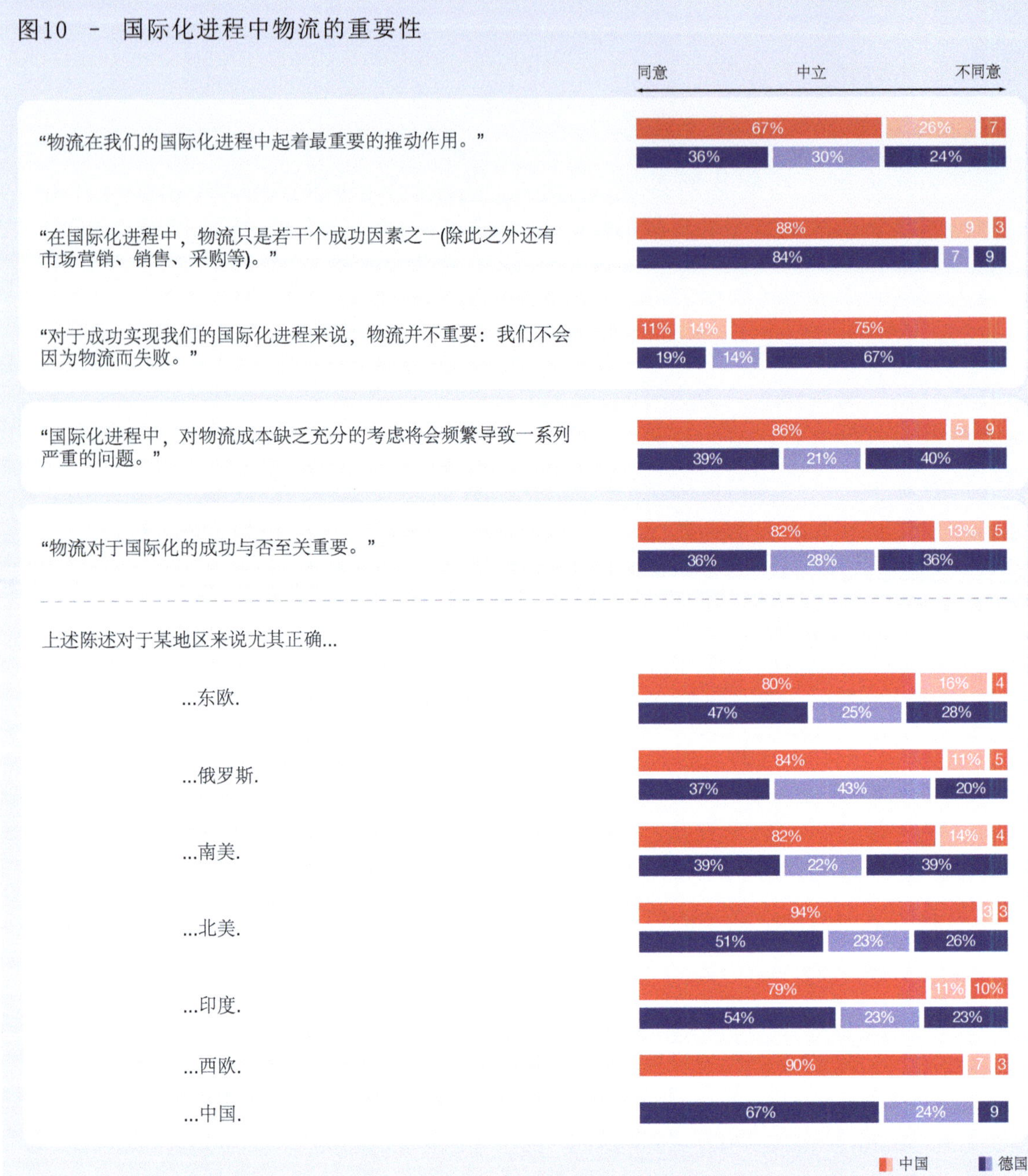

且会与公司的其它职能相互影响。物流可以被看作是一个"保健"因素，但是有获得巨大竞争优势的潜力。

86%的中国受访者以及大约40%的德国受访者由于在制定国际化计划时没有充分考虑物流，因此遇到了严重的问题。很明显，进入外国市场，物流是一个关键问题。高层管理者已经意识到了物流网络的

预算、资源分配以及战略规划对全球业务的开展十分重要。我们将在第5章中对此进行更深入的探讨。

3.3 国际化对物流的影响

就像物流促进了国际化的发展一样，国际化也直接影响着物流系统的特征。

国内和全球物流网络一个主要的区别就是距离，距离对物流的影响等同于运输速度以及可信度对物流的影响。在全球化的背景下，一项贸易所经过的环节增多，交付和补货的提前期就越长，同时需求预测的不可靠性以及实际产品的数量和质量的不确定性也会随之增大。因此，国际化直接改变了物流系统的某些关键绩效指标。在这一章中，这些改变将在企业参与全球竞争时所采取的策略中得以证明。

物流费用在总成本中的比例日益增大。由于某些原因，商业活动的国际化导致了物流费用在总成本中所占比例日益增大。调查显示，相对于中国企业，德国企业的物流成本增加得更多。这可以通过一个事实加以解释，那就是中国的物流条件还不够完善，使得物流费用占了中国企业总成本较大的部分。

除了运输对全球网络日益增大的影响，由于有时必须对产品进行地方性的调整（常被称为"本土化"）。这就导致了物流成本的增加，因为种类的增多意味着需要更多的管理，也降低了扩大经营规模的可能性。比起德国同行，中国管理者更加深谙这种关联性。德国企业多年前为了满足西方市场中顾客的个性化要求，就增加了产品的种类，而中国企业只在最近才开始这种尝试。相应的，约三分之二的中国企业看到了国际化的挑战，而在德国管理者中，这个比例仅为45%。

对库存水平的影响。由于国际化导致运输距离增长，超过三分之二的中国和德国企业正面临着库存水平的不断提高。

在企业进行国际化时，顾客需求波动增大且预测的准确性降低，也导致了库存水平的提高。这再一次成为挑战，而中国企业(70%)比德国企业(40%)更清楚地感受到了这种挑战，以买方为主导的西方市场在多年前就察觉到了这种现象，然而中国的企业只是最近才意识到了这一点。

为了优化库存，中国和德国的企业采取了不同的策略：

- 63%的中国企业和48%的德国企业试图通过集中管理库存的办法来降低库存水平。
- 74%的中国企业和36%的德国企业在最接近市场的地方建厂。
- 66%的中国企业和44%的德国企业致力于优化计划并提高预测的精度。

运输成本。由于运输距离增长，70%的受访企业都承担着运输成本的增加，在这一点上，中国和德国的企业没有区别。除了地域距离，国际网络中空运的比例也增加了，63%的中国企业和46%的德国企业同意该观点。在企业进行国际化的初级阶段时，就应该充分考虑成本计算问题，尤其是在设计商业案例时。不同行业的企业共同的经历显示，在全球网络中，空运在某种程度上说是不可避免的，对某些价

值较低的商品来说同样如此。意外的情况常常会发生，这只能通过空运得以克服，以确保交货期和关键客户的满意，或者只是为了避免重要的战略商品的缺货。

为了降低全球网络的运输费用，75%的中国企业和35%的德国企业试图优化路线来获得收益。许多物流服务提供商通常采用的方法，是把海用和空运结合起来，这既可以降低费用，又可以减少运输的时间。特别是在亚洲和欧洲之间的运输时，这种方法简单可行且效果显著。

另外一种方法则是结成采购联盟，它受到了中小型企业（SME）的广泛青睐，能够通过集中运输来获得较低的运输费率，从而降低运输成本。

物流网络的复杂性。进行全球化时，物流系统一个更微妙且具有挑战性的转型在于复杂性的增加。由于更多的合作者参与到网络中，70%的受访者经历了这种复杂性的增加。63%的中国企业和45%的德国企业强调了参与全球网络的物流服务提供商数量的增多。其中一个原因就是许多企业针对不同的运输模式，会选择不同的物流服务提供商，而国际运输本身就综合了海运、空运、铁路运输以及公路运输等各种运输手段。另外，当前很少有物流服务提供商能够提供真正的全球运输，即能够达到满足需要的服务水平和绩效。所以，企业趋向于在不同的地区或市场中选择不同的物流服务提供商进行合作，从而满足顾客的需求。

俄罗斯一览

自从苏联1991年解体以来，俄罗斯联邦一直在从国有计划经济向市场经济体制转型。这种过渡仍在进行，因此仅仅采取资源重新分配和重组现有的生产手段就可以实现巨大效益。

苏联解体后的开始几年呈现出经济持续萧条的景象，导致该国在1998年陷入金融危机。然而，俄罗斯的经济恢复迅速，主要得益于其最大的出口收入来源——原油和天然气市场价格的升高。到1999年，国内生产总值增长了6.4%，而2000年增长了10%。2006年，国内生产总值增长了6.7%，这个增长水平很可能持续下去。失业率在过去数年里一直下降，现在是7.2%，虽然通货膨胀率已经下降，但2006年仍高达9%。

俄罗斯联邦占地约17.1万平方公里（拥有142.8百万人口）是世界上最大的国家，覆盖了世界上九分之一的陆块。它拥有丰富的自然资源和世界上最大的天然气储量、第二大煤炭储量和第八大石油储量。[2]据估计，石油和天然气行业约占国内生产总值的20-25%。因此，俄罗斯经济直接依赖动荡的国际商品市场，这将带来极大的风险。在2005年，以燃料和原材料为主的出口造成了1180亿美元的顺差，石油和天然气占61%，而工业制成品只占所有出口的10%，制成品出口由于高估汇率遭受到高额的出口价格。

与2003年经济增长5251.8万美元相比，累计吸收外商直接投资仍然维持在较低水平。俄罗斯每年净流入的外国直接投资为30.461亿美元，只相当于中国的6.5%。[3]固定资本投资处于相对较低水平，占了18%，其中大部分都集中在主导的石油和天然气产业上。在"为俄罗斯联邦今后社会和经济发展的战略"（2003年10月）中，经济发展和贸易部强调俄罗斯的经济远离自然资源生产，以促进更广泛型增长的重要性。俄政府已经把运输和物流中心作为能够多元化的一个契机，希望把自己建立成为亚洲和欧洲运输的一个枢纽。

其他仅次于石油和天然气行业的重点行业包括快速消费品、高科技、电子、重工业、汽车和制药业。

另一个威胁到俄罗斯的持续发展和增长的是财富的高度集中。自1998年的危机以来，俄罗斯积极发展的经济果实，已经在区域经济发展中分配不均且产生了差距，这很可能阻碍经济持续增长以及消除贫困。

虽然贫困仍然是俄罗斯联邦的一个极大挑战，经济发展和整体上升的实际收入（10.7%，2006年）刺激了内需，顾客主要集中于满足被压制的需求，个人投资偏低。

当前的政治局势的特点是权力的累积，在中央和最高权威之间进行着转变。关注的焦点包括:国家控制的媒体、缺乏透明度、广泛的裙带关系和腐败现象，然而，政府在即将加入

世贸组织之前实施了几项改革措施。世贸组织明确规定成员必须通过立法程序，推动更多的透明度和独立的法律保护，以及进一步开放俄罗斯市场并减少官僚。这些改革包括金融和资本市场的改革，取消垄断(如电信和铁路部门)，制定反垄断法律并简化通关。尽管过去的几起事件反映出国家对市场主体发展的担忧，外商投资者仍然认为俄罗斯政治局势比较稳定。俄罗斯的商业氛围常常被形容为以连接和以好处为本的企业文化——即所谓的"喋喋不休"。

俄罗斯的自然环境也存在着挑战，如规模小、路途遥远、气候条件恶劣、基础设施（如电信网络、卫生和运输系统）正处于衰减状态。政府正在通过增加预算来解决这些问题，举例来说，俄罗斯正致力于对网络进行改善，如延长俄罗斯和中国的铁路连线。

俄罗斯的铁路网为不断发展的商业运输提供了坚实的基础，俄罗斯拥有世界第二大铁路系统，覆盖8.55万公里的铁轨。俄罗斯铁路持有大部分的主要线路，它是一家国有垄断企业，是俄罗斯的第二大公司，仅次于俄罗斯天然气工业股份公司。

铁路货运量约占俄罗斯装运货物总额的40%，如果不包括石油和天然气，铁路货运量约占装运货物总额的83%以上。铁路客运量在运输结构中占34%，是仅次于汽车的运输方式。

扩大来往于亚洲与西欧的跨西伯利亚的集装箱运输铁路，是目前最重要的铁路工程。俄罗斯铁路主席弗拉基米尔雅库宁指出，2006年俄罗斯的铁路面临的主要任务是，每半年为跨西伯利亚和贝加尔-阿穆尔主线增加国际运输量到100万个标准箱。这个项目包括与德国和中国铁路，以及兴建18号货柜码头的合约。俄罗斯铁路公司计划投资约43亿美元，从2006年到2008年横贯西伯利亚，项目还包括重新连接横贯西伯利亚至朝鲜的铁路以及持续检修网络。

整个九十年代，投资于交通基础设施的费用一直很低，直到最近才开始增加。然而从短期和中期发展前景来看，俄罗斯的基础设施预计不会得到很大的改善。

虽然基础设施和官僚主义是制约因素，物流市场在俄罗斯仍然得到了蓬勃发展，并奋力与物流需求相适应。2006年，俄罗斯物流服务市场的需求总量约为1200亿美元，近6000家公司从事物流经营业务，其中大多数提供运输服务。俄罗斯的经济正稳步增长，外国汽车制造商正在把部分生产能力转移给俄罗斯，与之相对应，提供更高级的物流服务成为了必须。因此，国际物流服务提供商意识到俄罗斯的物流市场存在着相当大的机遇，　所以他们纷纷在当地的基础设施上进行投资。

图11 － 在国际物流中跨国和各国具体的问题

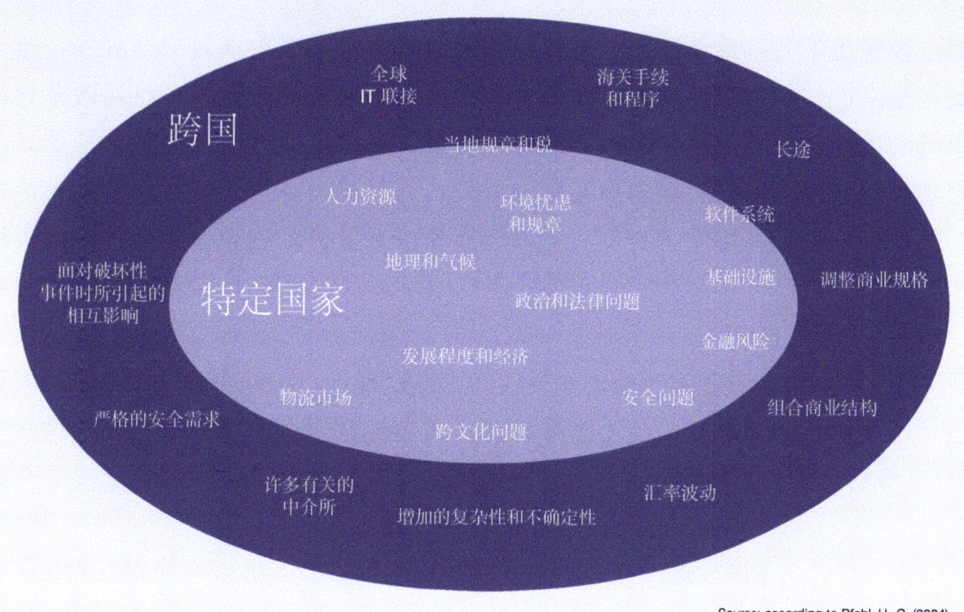

4 国际物流的挑战与困难

国际物流活动中面临的最大挑战就是多样性。每个国家和地区的物流系统都有各自不同的特点，从而导致各国物流系统有着很大的差别，但它们又必须在国际物流网络中加以协调。因此， 在设计和管理这样一个全球物流网络时，需要考虑到网络中各国家和地区之间所存在的一些差异，包括相互融合的商业和贸易结构，关注于运输时间和交货期柔性的同质化客户需求，运输基础设施的建立与管理，以及地理位置的分布等方面。

关于贸易和运输过程中所遵循的标准，只有一小部分是全球化标准，其余很大程度上还是依赖于各自的国家标准。

因此，在这种多样性的环境中实施全球物流策略，需要对企业所处的国内外物流环境有一个整体的了解，如图11所示。

另外，还有注意到两种不好的情况：首先，对于绝大多数国家而言，不仅其自身与别的国家不同，而且自己国家内部不同地区之间也会存在差异。比如在中国，人们也许就需要考虑沿海地区与内陆地区的不同。其次，由于许多环境都在不断发生变化，因此也使得全球物流环境也随之变化，而不是一成不变的。

4.1 全球化物流中各国的具体问题

全球物流中各国的具体问题包括很多方面。其中有些问题在不同的国家之间有着显著的不同，而有些问题则是在不同的国家和地区之间有着很大的相似性。

表2显示了不同国家和地区的物流经理们最近所面临的最严峻的挑战。

通过对德国和中国公司在不同领域内进行评估所发现的不同点进行分析，可以清楚了解到两件事情：首先，界定某一地区所面临挑战的严重性是一个认识问题，这种认识取决于该企业在国内市场所处的环境和该地区所面临的所有问题。例如，在北美，整体上的物流环境相对来说是很好的，因此硬件设施以外的其它方面的问题，例如不同文化间的问题，比较容易被重视。

第二，结果还显示，一些特定的模式是与当地社会和经济目前所处的具体情况息息相关的。下面将列出几个最重要的调查结果：

基础设施。对于所有的发展中的市场都是一个典型的问题，如东欧、俄罗斯、南美、印度、中国。很明显，基础设施的短缺，对经济整体的长远发展和增长都是个问题。从企业的角度来看，对基础设施短缺的认识对于成本控制非常重要 - 发展中国家与发达国家相比，前者的物流成本占总成本中的比例是后者的两倍。但是，在当地的竞争中，因为基础设施的情况对于这个市场所有竞争者都是相同的，所以也就显得不那么重要了。

安全问题。绝不仅限于物流方面，而应该放在社会和经济的大背景下理解。但是站在物流的角度看，安全问题是至关重要的。物流要运输所有有价值的货品，还要防止偷窃，抢劫和损毁，这些会导致低可靠性，长提前期和高成本（直接成本、间接成本）。在南美的部分地区，俄罗斯以及东欧，明显存在这方面的问题。当前由于多方面的原因，已经把东欧的部分国家列为运输中的"最危险"地区。

文化差异因素。以及这些问题的深度是和本国与目标市场之间文化的差异相关的。随着物流人员已经成为二者之间最重要的联系，跨文化管理在他们的日常工作中就显然变得重要起来。

下面将更加详细的描述各个国家的具体方面：

人员。人力资源的可获性和技能，在各个不同国家里也是显著不同的。非技术性人员，如卡车司机和仓库管理员等，在全球范围内都很容易找到。但是，受过良好教育的物流管理人才在很多国家则很难获得。这个问题在一个新兴市场里显得尤为突出。由于教育的缺乏和在快速发展阶段工人的短缺，任何行业的公司都面临着人员匮乏的情况。由于不同的国家对这些高级经理们的吸引力是完全不同的，所以关于人员的另外一个很重要的问题是委派人员到某个特定的市场去担任领导职位。一般来说，如果外派人员到吸引力较低的地区去，就需要支付更多的薪金进行激励。所以在早期计划阶段就要预见到对于培训当地员工和补偿外派的经理上要做

表2 – 各国具体的挑战问题

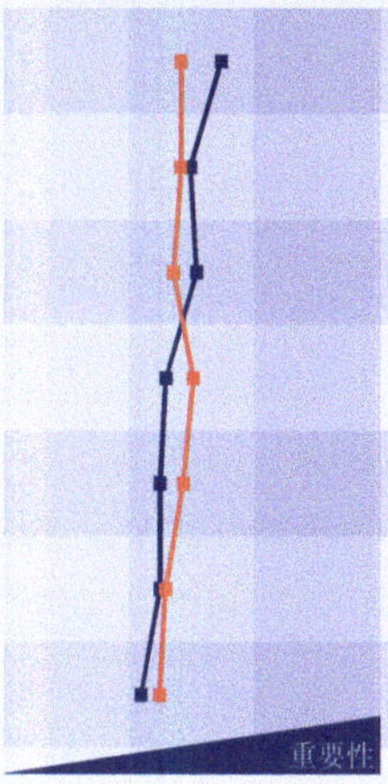

最多数的三个挑战和问题*	东欧	俄罗斯	南美	北美	印度	中国	西欧
跨文化问题	▲		▲	●▲	●▲	▲	●
政治和法律问题	●	●▲			●	▲	
基础设施	●▲	●▲	●		●▲	▲	
人力资源	▲	▲	●	●▲			●
安全问题	●▲	●▲	●▲		●		
物流市场				●▲	▲		
金融风险		●	▲	▲			●

重要性

* 列出最多数的三个回答。如果有多个并列第三，一起列出。

● 中国　　▲ 德国

哪些努力。

而如果是到经济更加发达的国家进行市场拓展，物流经理则必须面对这样一个问题：在这些国家，有能力的员工虽然容易获得，但是非技术性劳动力成本也相对较高。

文化差异问题。乍一看，我们可能会惊讶地发现，这个问题居然与物流有关。但是近几年来，物流学科的范畴得到不断拓宽，如今的物流经理必须应付日益增多的且传统上并不被视为物流领域内的问题＜参见 3.1章＞。文化多元化管理便是其中之一。研究表明，知识对于当今的物流管理者非常重要，因为与全世界不同民族的人们交往时，生活习惯、思想、语言、信仰、价值观、风俗习惯以及环境的不同构成了重大且关键的挑战。

从商业实践上看，文化因素必须从组织以及客户的角度来考虑。在这里，"组织"是指公司内部程序，例如各个国家的工作架构和完工效率。而从顾客的角度来看，则意味着了解当地的需求和顾客期望，对保证所需的服务水平和采用适当的处理方法是非常重要的。文化之间的明显差异对服务和货物的流动可能具有很大影响。例如在一些地方，守时被认为是一个非常重要的问题，而在其他一些国家也许并不是很重要。

在文化多元化背景下，不仅沟通和协调变得更加困难，而且对于物流从业人员的绩效考核和激励制度也开始向本土化转变。此外，这些差异是不断变化的，因为文化交流以及生活水平和教育在动态的全球环境中变得日趋相同。

因此，在不同国家进行企业管理运作时，文化被认为是独立且鲜明的。语言是不同文化之间最明显的差异之一，并在很大程度上影响了相互的理解。词句的意思常常要放在其被使用的语境中进行理解，因为语言是文化常用的一种表达手段，对于它所承载的特殊含义，不了解相关文化

背景的人就可能无法理解了。

全球网络中的另一个重要方面就是关于人们的宗教习俗、祈祷和祭祀的程序，我们也必须把它作为一个重要方面予以对待。 在对工作时间、假日、敬神地点等进行计划和组织时，尤其需要对这些问题考虑周全。

基础设施。由于货物运输的简易化程度已经成为决定一个国家经济实力的基本因素之一，所以国家的发展水平不同，其运输基础设施也会存在明显差异。

发达国家的基础设施十分先进和完善，不仅可以把全国的工业中心连成一个整体，而且也与其邻国紧密相连。目前发达国家面临的挑战是日益拥挤的交通.

在发展中国家，以下几个原因可能导致交通运输网络和设施的不足。经济的快速增长，以及随之而来的国内商品运输量的增加超越了现有基础设施的承载能力，这些现有的基础设施，通常只是为了满足基本的货物运输要求而进行建设的。此外，对于基础设施的发展的关注可能最初是由出口贸易所推动，所以仅能在发展的最初阶段确保贸易的增长和外汇流入。然而像今天我们所看到的，当企业为了发展而进入新市场以满足当地需求时，他们所依赖的国内系统的建立可能会因此而受阻。

例如，在几个主要经济中心之外的地区， 由于仓库和发运设施不能有效集成利用，同时也缺乏适当的基础设施，因此，尽管西部各省的制造和运作成本都相对较低， 企业也不愿意将业务外包给这些地区。

交通线路的质量，并不完全取决于诸如铺设不平坦、渗水这样的路面条件，还取决于整个交通网络的便利性和分布密度。在有些国家，道路网络集成率相当低，道路之间联系很少，公路和跨线桥几乎不存在。常见的情况是，国内的基础设施零碎分散，而且包含很多质量和维护水平之间的地域性差异。还有，其他的一些因素也可能导致交通堵塞、安全事故和车速降低，从而增加运输时间以及相关的成本。这些因素包括，不良的交通控制系统，错误的交通标识，交通意识的缺乏，以及在路面上行驶的一些低速交通工具，比如拖拉机和自行车。举个例子，在印度，公路上的日平均可行里程是250公里，然而在发达国家这个数字都在600公里以上。

物流市场。近十年以来，物流服务的外包是一项降低成本，提高服务质量的有效方法＜参见 5.6章＞。在全球一体化过程中，它在更广阔的范围内，发挥了更重要的作用。

可是，在一些特殊的市场，物流的服务和质量正在发生着很大的变化，国际物流服务提供商直到今天也无法制定出一个所有国家和地区的统一标准。比如在中国，直到2005年底，仓储业务对于国外的投资都是不开放的。由于缺乏投资，在中国一些简陋的仓库里，天花板过低、照明不足、港口的缺乏以及不正确的保护措施，这些都是导致货品破坏率和损坏率高居不下的原因。据估算，中国每年有25%的水果和蔬菜因为保存不当而腐烂毁坏。而且不同国家所采用的不同的货品转移方

法也可能使国内货物的运作出现混乱。

例如在印度，内陆集装箱仓库（ICD）大多紧邻工业中心和大城市，以便于海关更快地甄别和分配货品，从而避免在关口造成堵塞。因此，通过主要的铁路交通线路，进口集装箱就能被传送到这些免税区。也正因为如此，印度国内的交通运输网络管理需要极大地依赖于政府基础设施建设。

我们可以从第五章看到更多详细的情况，进入新市场之前对物流服务商的全面调查分析是至关重要的，并且制定一个清晰的如何与有能力的物流服务商合作的战略是物流系统国际化成功的关键。

政治和法律问题。国际物流运作必须遵循国际和本国的规章制度。调查显示在东欧、俄罗斯、印度和中国这样的新兴市场里，政治问题明显更具挑战。大多数新兴国家制度不完善，导致非正式关系具有很高的重要性。在新市场开始运作时面临最重要的方面是：贪污和裙带关系问题，与当地权力部门进行非正式接触的重要性，政府权力部门独断专行，制度的完善性和政治的稳定性。

尤其是对于德国企业，经常会低估与权力部门甚至当地供应商进行非正式接触的重要性。由于在许多地区都没有像西欧这么重视书面合同和法律，因此在任何国际化活动的初期，就必须考虑到这方面的问题。

近几年，许多国家在促进和简化商品运输方面的规章上，尤其是在进出口的运作方面，做出了巨大的努力（例如：物流服务市场的自由化）。但是不论是在发达国家，还是在发展中国家，依然存在过度管制的问题。在这方面俄罗斯做出的改善最为显著，从2004年开始，俄罗斯已将海关条例从3500种减到了439种。

财务风险。由于物流人员手中往往掌握着大量的仓储库存设备，因此这其中的财务风险管理近些年已经受到重视。在全球化运作下，企业在各个不同的国家里，面临着各种不同的财务风险(如阿根廷和亚洲金融危机)。金融危机的经验更说明了财务风险管理的重要性，并且物流或者供应链经理们在设计物流网络、补货策略等的同时必须关注财务风险。目前，欧洲和中国的物流经理们认为这个问题在俄罗斯和北美、南美最为严峻。某些国家的财产权利仍然是令人关切的，但最重要的是货币汇率的波动。采用套期保值策略与优化的付款和配送，企业可以设法减少风险，以减轻各自的后果。

安全问题。虽然很多公司主要着重于提高效率，但安全问题已成为另一项重点关切的问题，尤其是在全球物流中。这个问题，当然不是仅限于国家或地区界限内。然而，调查显示，从物流人员的角度看，不同的区域内对问题的评估结果是不一样的。安全问题不仅需要考虑犯罪行为，如盗窃，抢劫和欺诈行为，而且需要考虑那些越来越具有政治动机的攻击，这可能是针对贸易渠道的。今天犯罪分子的目标已经不仅是容易被窃取的高价值消费品，而且其他中等价值产品和不同类型的工业货物也成为了他们的目标。针对大宗商品的全球有组织犯罪的增长，正说明了

全球范围内货品损失增加的原因。

　　当我们对保护国内货品免受盗窃和抢劫的挑战进行观察时，每个国家的个人安全标准变得显著起来。在世界的某些地区，糟糕的经济环境和有组织的犯罪，使得储存和运输货品的危险增加。抢劫当然会对货品带来危险，但是同时也会对司机和员工的安全带来威胁，因此十分有必要采取措施来加以避免。在某些国家，一些物流服务提供商对运输卡车进行护卫，委派武装保安人员在内陆运输中随行。在东欧，尽管水路运输要比陆路运输慢上三天，且成本也更高一些，但是为了避免高价值货品的被盗危险，很多运输商还是对从德国到俄罗斯的货品采用水路运输。

西欧一览

　　虽然西欧这个词主要用于冷战时期，但是在1990年冷战结束后还是沿用了这个叫法，至今已经17年了。在最近的一次调查中，西欧包括的国家有奥地利，比利时，丹麦，芬兰，法国，德国，希腊，爱尔兰，意大利，卢森堡，荷兰，挪威，葡萄牙，西班牙，瑞典，瑞士，还有英国。这些国家中除了挪威和瑞士之外，在2004年欧盟进行扩展之前，就已经是欧盟（欧盟15国）的成员国了。

　　西欧地区的人口总数只占全世界人口的6%，但是该地区的GDP却占据了全球GDP的24.5%的份额。在这次调研中所涉及的所有国家和地区中，西欧以3970亿美元轻而易举地成为最大的FDI，是第二大FDI（北美地区）的2.5倍。该地区的大多数国家都可以归为发达国家。西欧的经济特点在于高份额的人均GDP值和先进的科学技术，同时在2006年的世界财富500强名单中，西欧企业占据了35.4%的份额，这也充分说明了这一点。

　　由于欧盟的东进，就物流方面而言，德国变得越来越重要。德国现在处于各国市场通道的正中心。它是欧洲大陆的贸易中心，横贯东西，直通南北。到2006年止，以高度发达的经济和高质量的生活水平为特征，德国成为了欧洲第一、世界第三的经济体，排在美国和日本之后。还有，德国也是全球最大的出口国以及第二大进口国。物流业是德国竞争力中的主要支柱之一，能够为产业增值、货物流通和企业协作铺平道路。排在汽车工业和零售业之后，物流业是德国的第三大行业，从业人员有260万。物流行业的发展速度也远远高于各行业的平均水平。这不仅是由于德国的地理位置处于欧洲正中心，而且在国外投资者的眼中，德国完善的基础设施和处于国际顶尖水平的物流技术都是使得物流行业高速发展的重要原因。

　　由于东欧的加入，欧盟的扩张导致了物流节点的重新定位。而这些节点周边所存在的潜在商业机会造就了一批新兴市场。这使得以全面的"泛欧"市场为目标的物流结构发生了变化。另外，配送中心也由"三国经济联盟"（比利时，荷兰，卢森堡）转到了东欧，例如波兰，因为其国内的工资水平相对于西欧地区而言更加低廉。不过，整个欧洲的货物供应，无论是来自亚洲还是北美，绝大部分还是通过西欧转运的，特别是在鹿特丹，汉堡，安特卫普，不来梅等港口，也就是所谓的"西北圈"。

　　欧盟的第二大积极作用就是协调欧洲的运输基础设施。相对于世界的平均水平而言，西欧的公路、铁路还有机场高度发达，这对物流服务行业是十分重要的。

　　由于"西北圈"（鹿特丹，汉堡，安特卫普，不来梅）的规模大，功能全，它们在北海港口中地位很突出，可被称作多用途港口。就货运处理量来说，鹿特丹是欧洲最大的港口，而且在货物总量上排第二位。在集装箱运输方面，鹿特丹，汉堡和安特卫普也是位列前10位的港口。在欧洲，集装箱处理总量高达6100万标准箱。不仅是"西北圈"的港口有发展潜力，地中海沿岸的港口也保持着9%的年增长率，如意大利的热亚陶罗和热那亚，西

班牙的巴伦西亚和巴塞罗那，希腊的比雷埃夫斯。需要指出的是，鉴于各主要集装箱港口多为转运港口，为内陆需求服务，西欧4%的标准集中箱用于该地区的内部运输。

西欧也是世界上最重要的一些机场的总部。西欧有四大主要机场：德国的法兰克福，法国的巴黎，英国的伦敦以及荷兰的阿姆斯特丹，这些机场在地理位置上都相互毗邻。它们的运输能力总和占到该地区空运总量（540万吨）的67%，因而它们肩负着欧洲干线的集配功能，专注于国际空运。例如，在法兰克福，亚洲市场的空运量为771,000吨，占总空运量的49%。相反的伦敦则专注于北美的贸易运输往来。

总而言之，西欧从现代物流理念的高度扩散中获益良多，因为这是一个供不应求的行业，而且物流市场相当发达。到目前为止，24小时不间断运输配送在西欧已经成为基本服务准则，而在东欧则暂时尚未实现这一点。展望未来，合约物流将同全球代理物流服务提供商一样在全球化进程中扮演重要角色。迄今为止，西欧的外包物流服务程度仍然高于世界其它国家和地区（如中国）。

在西欧，技术和环境标准备受关注。例如，ISO14001环境管理标准的出台就是为了帮助企业减小其行为对环境的负面影响。这样的生态标准已经高度渗入到西欧的物流业中。例如，为了加强物流行业的可持续性发展，政府利用行政手段对道路进行收费定价，以支持和发展生态型铁路的修建。可以想象，这项干预措施对东西欧的路网配流有着长远的影响。

随着欧盟的继续扩张，东西欧间的进一步合作值得期待。而随着合作的进一步加深，东西欧间由于历史政治形态而导致的分歧与障碍会逐渐减弱直至消除，使得欧洲最终成为一个联合的经济和政治体，这也是更加令人兴奋的原因所在。

图12 - 在全球物流网络中面临的跨国问题

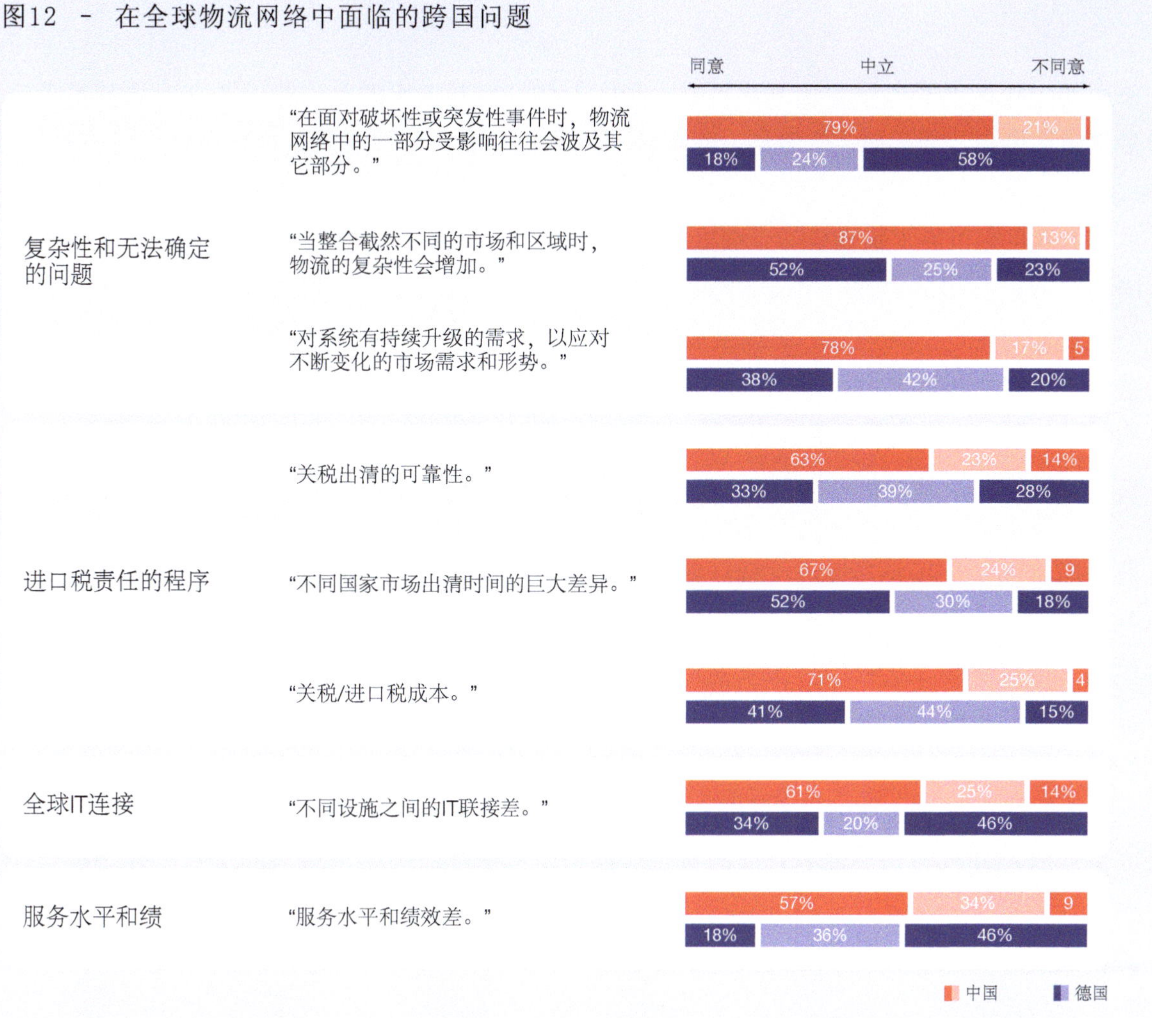

4.2 国际物流中的跨国问题

在建立国际物流结构时，除了要考虑国内的特殊问题外，还要考虑大量有可能存在的跨国问题<参见图12>。

复杂性和不确定性。 由于需要管理和处理很多的商品和信息的跨国传输，从而导致国际物流成为了一个相当复杂的问题。漫长的运输距离、易出差错的信息交换、繁杂的国内外交接手续、日益增加的运输成本、风险以及更长的提前期，这些因素都构成了国际物流的复杂性。在国际物流网络中，信息交换和物品传输中的失误所带来的风险是很大的。出现这种问题的主要原因是在对于这些国际物流活动中的计划、销售和控制中，参杂了大量的人、公司和机构的活动。

这些中介物可以分成几类，例如海关、管理部门等权力机构；工会关税和航运团体；航空公司和航运公司等运输公司；货运和海关经纪人等物流协调部门以及收货人和发货人这样的客商。除了繁杂的文件手续过程之外，还要在不同的中介之间搭建"桥梁"，这不仅仅是对于在实物处理上的挑战，更是对于信息数据正确传递的挑战。由于各个部门信息系统的区别，信息很可能出现错误、缺失甚至在传递过程中出现丢失。显而易见，对于国际物流网络的管理必须建立在更多的合作与沟通的基础上，以及对于监控系统的持续

改进之上。

供应链中断时的相互依赖。在当今的的国际价值链中，各方面成员间存在着千丝万缕的联系，因此价值链中某个某一环节上出现的微小中断，可能会造成及其严重的后果。约有80%中国企业已经看到了这个问题的严重性，而德国的企业有信心通过风险管理降低发生概率，所以在德国只有少于20%的企业把这个看成严重的问题。

但是我们必须看到，近年来供应链出现中断的几率增大了。近三十年来，世界范围内的技术性的灾难，也可以被称作人为的或运输的事故，如船只的丢失，火车相撞，以及油库的爆炸等等，已经呈指数增长趋势，这些都增加了供应链中断的几率。与此同时，自然灾难，如：地震，海啸，火山喷发，泥石流以及狂风和龙卷风，也发生得越发频繁。结果，这不仅造成了人员的巨大伤亡，同时也破坏了基础设施。这将导致暂时的混乱，甚至有可能造成全球供应链的整体中断，给受灾地区的企业带来长期而深远的影响。

例如，2003年严重急性呼吸道综合症（SARS）在东南地区的内陆和东亚的爆发导致了地方制造的下降和口岸的关闭。因此，由于缺乏进口的部件，美国的计算机装配生产也不得不临时关闭。

进口税费和通关程序。另一个跨国问题是贸易壁垒的结果，可以分为关税和非关税措施。税是指各种类型的关税，而非关税措施是指直接和间接的贸易保护主义方法，譬如进口限制。由于海关法、关税和过关手续因国家而异，　这就增加

了复杂性，并且必须对各个国家和区域的规格分别进行分析。从跨国的角度看，这个调查表明，除了关税带来的直接费用以外，过关手续所耗费的时间也几乎同样重要。在许多地区，过关手续中可靠性和透明度的缺乏影响了全球性运输的可预测性。因此物流人员通过附加的安全库存和额外的管理来应对，从而可能极大程度地影响国际经营活动的整体成本结构的间接费用的发生。因此，当我们考虑走向全球的时候，这个问题不可以被低估。

全球IT网络。在全球物流中，准确性是极重要的，由于我们提到的价值链中各成员间复杂的相互依赖性，在某一特定阶段发生的错误和发货的延迟，在价值链上就可能被放大至最后的顾客。因此，加强使用先进信息技术来制定计划并管理物流，对全球物流至关重要，包括运输途中的商品以及库存和装货调动点的商品的可见性。用技术术语说，它允许公司关注于物品的状态并战略地处理供应链。如果对供应链进行端对端地整体观察，他们能根据信息进行活动并预见这些行为将如何冲击转载货物的运动。显然，货物在运送途中具备可见性，能够为托运人带来更多好处，但同时也是全球物流中最富挑战的有待完善的功能。跟踪系统必须通过各种手段网罗包含大量数据的中介，包括网络服务，EDI和遗留系统等。

但是不同国家不同公司以及同一公司不同情况下的IT前景都是大不相同的。近年来，全球供应链上供应方和需求方的IT联系成为一个主要的问题。这次调查显示，1/3的被调查的德国公司和接近2/3的中国供应商因为他们精糕的全球设施的IT整合而感到苦恼。

尽管技术上已经有较大的进步，　例如，新的中间产品和新格式数据交换，但这个问题将依然是未来全球网络的关键问题。

5 进入一个新市场的物流策略

本章节分两步分析了进入一个国外市场的过程。首先，从国际化进程中物流的视角来观察如何在国际化进程中整合物流的。然后分析了影响物流策略成功的关键因素，从而为在国外市场设厂的策略提供指导。

图13简要地描述了一些问题，我们会在后面章节中对这些问题进行详细的探讨，此图还为读者了解分析结构提供了帮助。对即将迈向国际化企业的物流管理而言，此处提到的六个不同的问题彼此紧密相关，且非常重要。图13中的红线和蓝线显示了中国和德国的大部分成功的企业在这些项目上的情况。后面的几个子章节基于完整的调查样本做了详细的分析，图13仅仅显示了从大部分成功的公司中获得的数据。这些公司在它们涉足的50%以上的国外市场上完成了预先设定的物流目标。

国际化过程时期〈参见5.1章〉。我们提出了一个连续步骤的流程模型，以时间为轴线分析国际化活动，从最初的设想到实际开展国外业务，再到新的物流系统的实施和在全球网络中的整合。调查结果显示，中国和德国的物流经理们都面临着非常严峻的时间压力。比较而言，中国公司的物流或许要更快一些。绝大部分的成功公司在国外开设工厂的速度都超过了平均水平。

物流经理的参与〈参见5.2章〉。调查显示，从国际化进程伊始，物流经理就深入参与其中，从而在全球网络中获得更高的成本/时间效率，并获得更多的竞争优势。从两国公司的比较中可以发现，中国公司更加注重考虑物流方面的问题，而且他们物流人员的参与程度要更深一些。而在德国公司，物流经理们的影响则相对比较有限。

构造国际化程序〈参见5.3章〉。中国和德国的公司都认为，国际化程序和重复过程的标准化程度越高越有益。相应地大部分成功的公司都有类似的看法，包括标准化的市场分析步骤和指导方针等等，这些程序提高了资金使用的效率和进入新市场的速度，也增强了最终系统的性能。从所有权的总成本等问题的国际化路线设计和应用上而言，中国和德国的公司有着很大的区别。从几十年前起，德国公司就已经渐渐走向了国际市场。因此德国公司的网络随着时间的推移逐步地发展，但缺少全球性的战略规划。与之相比，中国公司现在正处于全球扩张的初始阶段。他们中那些成功的企业都遵循了Greenfield的方式，他们着眼全局，制定了复杂的计划来正确地部署全球物流网络。

第5章的前三节论述了国际化进程的特性，随后讨论了成功进入国外市场的物流策略的三个关键因素。

如何应对变化的环境〈参见5.4章〉。由于企业面对着未知的市场需求，面临着不断变化的商业环境，因此，柔性在国际物流网络中的重要性变得日益突出。基于此，我们分析了在企业迈向国际化的过程中，物流战略的柔性应该达到何种程度。大多数成功公司案例的表明，中国公司比德国公司要更注重柔性。德国

图13 - 国际化的框架

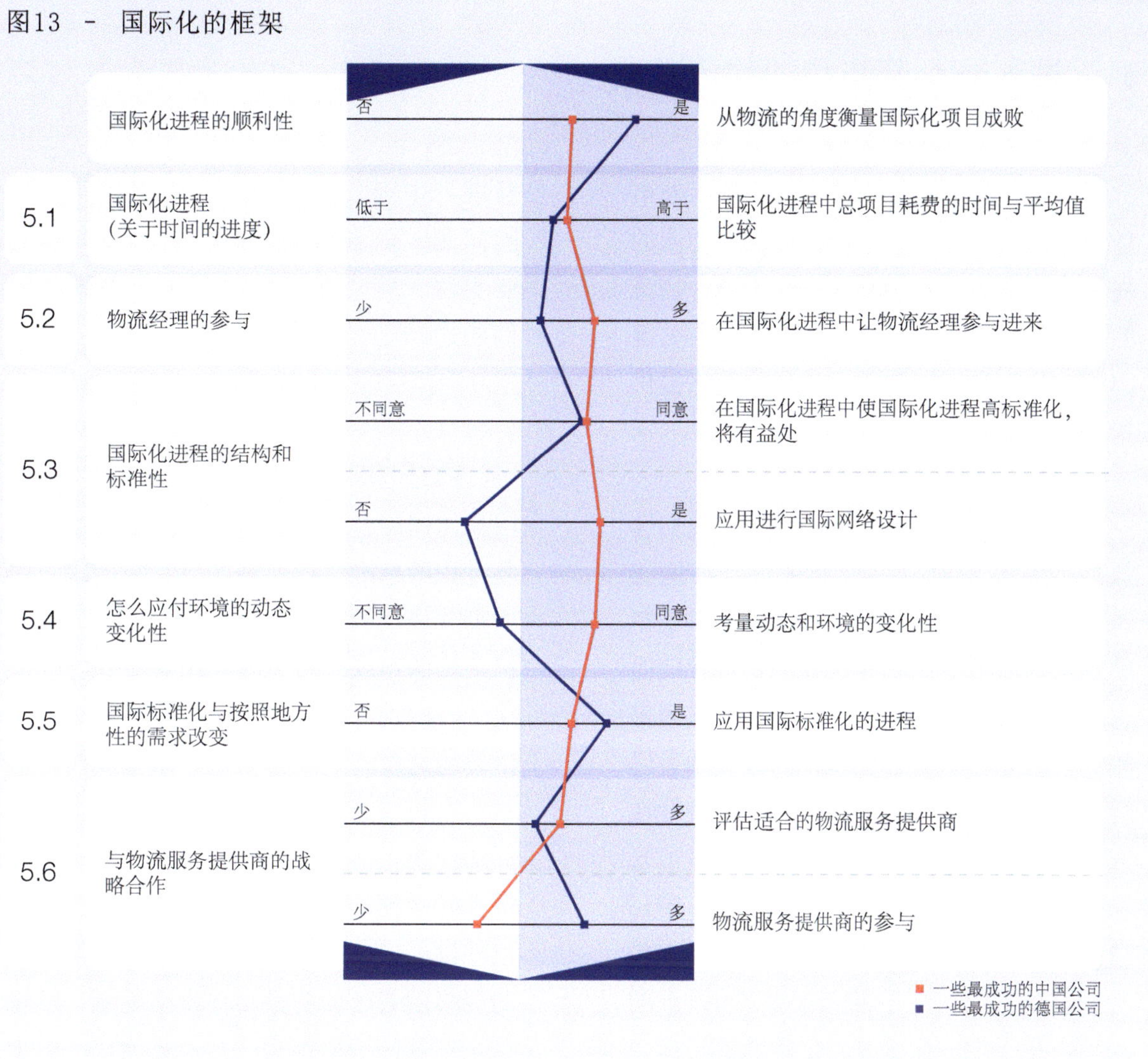

公司则更注重系统的成本、速度等数量指标。

全球化过程的标准化〈参见5.5章〉。在国际物流系统的设计过程中，为了减少复杂性和适应当地情况，满足当地特殊的物流环境和客户需求，必须介于全球标准化的两个极端之间，制定合适的决策。虽然应用全球化的标准流程有着很大的局限性，但是大多数被调查者都还是认为它是有益的。在那些获得成功的公司中，全球标准化程度都高于平均水平，对于德国公司来说更是如此。德国的成功公司与一般公司之间的差距相当之大。

与物流服务提供商的战略合作〈参见5.6章〉。当企业进入国外市场后，往往会采用更多的外包策略，企业把越来越多的物流活动外包给物流服务提供商。中国和德国的被调查者都认为与物流服务提供商合作是有益的。在全球物流管理中，物流外包被认为是个战略问题。尽管大部分成功的中国企业与德国企业相比，更重视与物流服务提供商的合作，但是他们最终只是将较少的部分业务外包了出去。这可能主要是因为在中国外包业务的发展还不像在西方国家那样广泛。因此，当全球竞争者加入并改变业务外包的运作模式之后，中国的公司也会在未来将更多的业务外包出去。

我们将在后面的几节里对这些方面进行详细论述，图13对整个第5章的内容作了总览。

东欧一览

1989年冷战和铁幕政策瓦解之后，东欧开始进行大力改革，从计划经济模式向市场经济模式转型。本次调查中涉及的东欧各国是保加利亚、克罗地亚、捷克斯洛伐克共和国、爱沙尼亚、匈牙利、拉脱维亚、立陶宛、摩尔多瓦共和国、波兰、罗马尼亚、斯洛伐克、斯洛文尼亚和乌克兰。

2006年，东欧人口总数和GDP分别占世界人口和GDP的2.4%和1.5%，与此同时，西欧的相应比例分别为6.0%和24.5%。东欧近几年来吸收了大量的外来投资，其中部分原因就是因为它是一个经济快速发展的地区。东欧经济具有GDP持续增长、工业产出增加和失业率下降的特征。一些东欧国家为跟上现代科技发展脚步，实现现代化生活方式做出了巨大的努力。例如在爱沙尼亚，部长们利用手提电脑进行无纸化会议，大部分居民都采用电子邮件的方式进行纳税，然而，在整个东欧范围内，还是存在着巨大的反差。墙壁斑驳、道路崎岖的破败村庄与新兴繁荣的摩天大楼比邻而居。在波兰的小村庄里，近一半的人没有工作。前东盟的其它一些国家也到处可见破旧的房屋，时间似乎还停留在那个时期。在过去几年中，贫富差距越来越明显。

2004年欧盟进行了扩张，将捷克斯洛伐克、爱沙尼亚、匈牙利、拉脱维亚、立陶宛、波兰、斯洛伐克和斯洛文尼亚等一些国家吸收进来。同年，克罗地亚也进入了候选之列，摩尔多瓦和乌克兰也有希望加入其中。2007年，保加利亚和罗马尼亚成为了欧盟成员国。如果我们将东欧与西欧进行对比，可以发现他们之间存在着明显的差异，不仅反映在文化和语言方面，也体现在经济和工业的优劣势上。近年来一些西欧企业形成了一个新的商业理念，即在东欧国家进行"近岸生产"，以发挥其成本优势。随着东欧国家不断加入到欧盟中来，这种优势显得越发明显，西方企业获得了大量的廉价劳动力。总而言之，东欧的历史已经发生了很大的改变，现在东欧已经包容了多种语言和文化，成为一个多元化地区。随着欧盟的进一步扩大，各国间的历史隔阂将逐渐淡化，整个欧洲市场，包括运输和物流设施，将会共同和谐发展。

东欧的铁路经过几十年的发展，如今已成为最重要的运输工具，遍布的铁路网络为地区性的物流运输服务提供了便利。从西欧的发展过程来看，我们可以预期到未来的运输发展将会由铁路逐渐转到公路上。近海运输方式对于东欧企业来说也是相当重要的。尽管东欧的港口在世界上的地位无足轻重，但近海运输仍然是常用的装卸发运方式，波罗的海港口也能从欧洲的海洋运输中获利。不过，与西欧相比，东欧的公路、铁路、机场建设仍然处于不发达状态。

在进行改革之前，东欧的物流企业大多是国有企业，体制僵硬，作风官僚，并实行计划经济模式。随着经济的发展变化，正如中国改革开放以后的情形一样，现在物流市场已经涌现出许许多多中小物流企业。事实上，没有任何一家欧洲物流服务提供商能够依靠自己的实力为整个欧洲物流服务，即使是西欧国家的物流企业也是如此。因此，未来泛欧物

流网络的扩张将大有可为。东欧之所以没有全面持续地进行基础设施建设，一个重要的原因是各成员国之间的异质性。不过，近几年建立的横跨东西欧各国的所谓"泛欧洲运输长廊"，使得各个国家和地区间的交通运输更加便利。目前，主要的困难是筹集发展资金，尽管欧盟承诺会进行资助，但是主要的资金仍然是由成员国各自承担。

在发展国际物流服务的过程中，西欧的影响仍然是相当显著的。例如，从亚洲和北美运来的货物和商品在通过火车、卡车和飞机运往东欧国家之前，首先会运至西欧港口如鹿特丹、汉堡、安特卫普、布莱梅，即著名的西北地区。同样，法兰克福、伦敦、巴黎、阿姆斯特丹等几个主要货运机场也对东欧的供应非常重要。所以，几乎所有运至东欧的货品都是取道西欧的。

随着东部扩张中政治和经济的转变，新兴的高度发达的物流基地对物流发展所起的作用越来越重要。与西北地区已有的一些重要的港口的重要性相比，许多大型的配送中心建立在东欧并且会在将来有很大影响。布拉迪斯拉法（斯洛伐克）、康斯坦萨（罗马尼亚）、敖德萨（乌克兰）以及波兰的卡托维兹将起到关键作用。拿波兰来说，总人口超过三千八百万，是新加入的欧盟最大的成员国，它拥有产能巨大的生产基地和大规模的顾客群体。另外，考虑到时间和成本的节省，尤其是考虑到欧亚大陆之间的连接作用，黑海的港口和跨西伯利亚铁路将会适合于国际运输。

毋庸置疑，差异性影响着物流市场。物流服务的质量和绩效在东欧国家中是有差别的。一方面，拥有着全球化物流提供商的先进物流市场已经存在，另一方面，落后的物流结构也并不少见。从物流系统的结构来看，波兰、斯洛伐克、匈牙利、和捷克斯洛伐克共和国属于发达经济体。直到上世纪90年代末，斯洛伐克还因为拒绝成为欧盟成员国而被视为东欧难治理的国家，而现如今，已被国外金融投资家视为投资宝地。例如，著名的国际投资商起亚集团于2006年下半年在斯洛伐克建厂，很大程度上带动了该国汽车工业的发展。这个工程预计将会提供1万个新的工作岗位。起亚集团并不是第一个在偏远的东欧国家建厂的汽车生产商。德国大众公司在布拉迪斯拉发建立了其豪华车–雷克萨斯SUV的生产基地。因此，斯洛伐克凭借本国廉价的劳动力和低税收成为欧盟最吸引外国投资者的国家之一。另一方面，匈牙利和罗马尼亚则由于其地理环境和工业结构，而大大制约了物流业的发展。尽管由于东欧地区的大量廉价劳动力促进了该地区的物流发展，不过也应当看到，地处欧洲和俄罗斯之间的波罗的海国家也越来越显现出他们成为重要的物流基地的潜力。

5.1 国际化过程的持续时间

现在全球市场的变化速度日益加快，时间也变得越来越重要。事实上，在进入一个国际市场时，时机的选择被看成是商业决策不可或缺的一部分。因此"时间"包括了两层意思：一是指进入市场的时机，二是指完全融入国外市场的时间长度。为了满足时效的要求，降低进入市场的成本并充分利用资产，时间问题给国际化带来了很大的压力。中国企业认为他们的国际化过程需要14个月，而德国的企业则认为平均需要18个月。

国际化过程的组成。在这次调查中，国际化过程指的是从刚进入国外市场时形成初始的战略思想，直到完全融入一个新的物流系统以及和全球网络中的所有公司结成一个整体。大致上讲，在大多数企业的决策中，这样一个过程包括几个步骤。这可以被看作一个组织严密的过程。在刚开始时，最高层做出决策，然后依次灌输到中层和低层，并开展相应的运作活动。

图14显示的是一个简化过的国际化过程事例。它包括了四个连续的过程，从物流的角度描绘了进入一个典型国外市场时的主要阶段，即：（1）对新市场进行评估，（2）制定战略计划，（3）制定物流计划，（4）实施物流系统。

当然，在实际情况中，这要复杂得多，而且各个企业的情况也不一样。但这个简化模型在研究物流系统国际化的时效性方面，还是具有价值的。

步骤一：对新市场的研究。在一个新的市场或地区发展时，国际化过程的

第一步是从各个方面进行市场分析。与此同时，最高层需要对目标市场的所有条件进行系统评估。譬如：销售潜力（市场的大小以及潜在的市场份额）、进入的要素（如当地的供应商基础、劳动力成本以及资源）、国内以及国外运输的基础设施、相关政策及法律等等。我们对进入市场的目标集中进行了详细介绍〈参见2章〉，也对可预料的不同商业案例做了研究。如果研究显示市场进入是可行的，且能满足预期目标，那么就可以做出进入市场的决策，第一阶段告一段落。

对一些公司来说，最初的想法来源于内部KPI以及国家特殊政策的指导，从中获悉进入哪个市场会比较有利的信号。

对于大多数德国和中国的企业来说，这个阶段占到了三分之一。正因为要考虑这么多方面，所以需要专家的参与以及专业的技能，才能构造出全面而现实的市场远景。

总的来说，德国和中国的企业在这个阶段大概需要5~6个月时间。相对来说，中国企业的速度似乎要快一些。

步骤二：制定战略计划。做出了进入市场的最后决策之后，下一步就是制定战略计划了。在该阶段中，最高管理层确定了进入外国市场的大致计划、时间表、各个阶段的里程碑以及财政预算和各部门的具体目标。明晰了计划中跨部门的合作事宜，建立具体的措施以保证杜绝意外的产生。

该阶段的结果是形成了一个框架，该

图14 － 国际化进程中项目的阶段

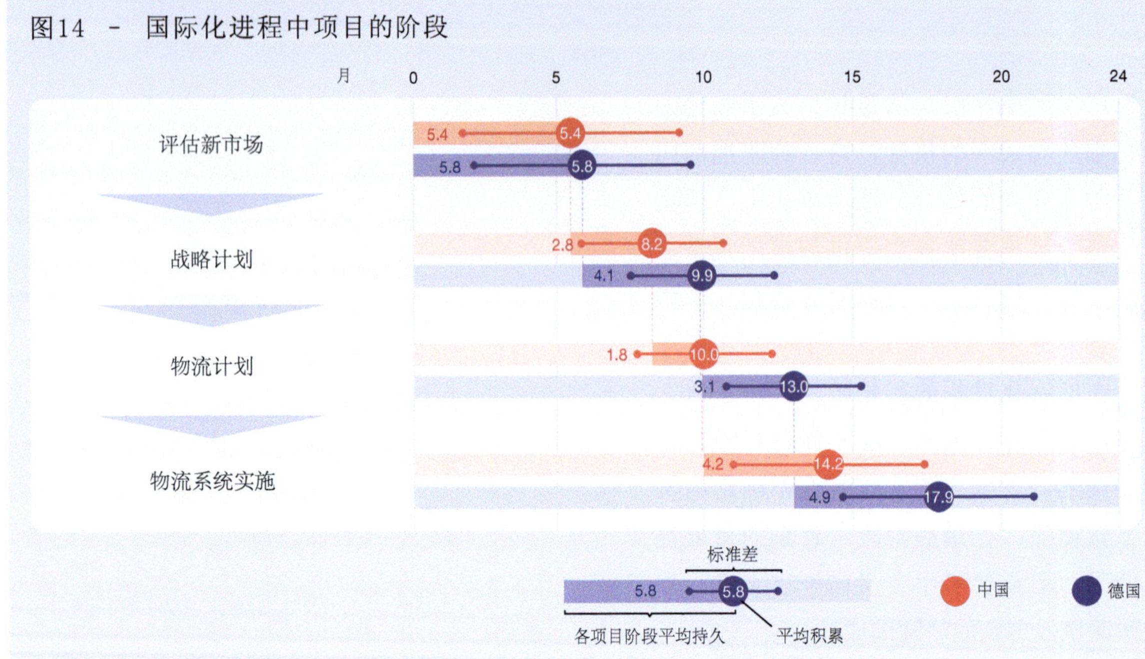

框架可以指导企业的各个职能部门根据企业的战略目标制定相应的行动方案。

相对于第一阶段，该阶段要短得多。中国企业最多只需要3个月，而德国企业也只需要4个月多一点。

步骤三：制定物流计划。当制定了战略计划后，不同的职能部门开始制定明确的计划，这其中就包括物流部门。在和其它相关职能的相互作用中，物流的过程得到了明确，建立了相应的执行计划。与物流计划相关的方面包括确定服务水平、提前期、库存策略、网络结构、容量的计算、设施的分配（仓库和交驳）、IT集成以及物流外包的决策等。

相对于其它两个阶段，该阶段要短一些。在中国的企业中，该阶段不足2个月，而在德国的企业中，该阶段也只有3个月多一点。这证明了复杂的物流结构是在很大的时间压力下建立起来的。

步骤四：物流系统的实施。大约在进入全球市场的10个月后，中国的企业开始实施新的物流系统。而德国的企业还需要多等3个月。

物流系统的实施对于国际化的成功十分重要，也相对更耗时一些。在中国，该阶段大概需要4.2个月，而德国的企业则需要4.9个月。该步骤的成功与否能够验证前期做的计划和决定是否真的和现实条件相吻合。该阶段属于操作级别的，需要系统具备解决问题的能力。在全球范围内对所有过程，以及人员和技术进行整合具有很大的挑战。企业必须对员工进行训练，不仅包括物流方面的，同时也包括领导能力以及跨文化的管理能力，而且还要给员工授予必要的权力。

国际化过程的特点。一般来说，中国企业的国际化过程大概是14个月，而在德国这个过程大概需要18个月。就像在图13里描述的那样，成功企业的过程耗时低于平均值。相对于其它企业，这些成功的企业能够更有效地应付时间压力所带来的挑战。

毕竟，整个国际化过程都致力于从最

高管理层到操作层，从最初形成国际化的最初想法到最终的执行，进行跨职能部门甚至跨企业间的整合。然而事实上，现实中的整个过程要比我们提供的模型复杂得多。在特殊情况下，可以同时进行几个步骤，而且不能忽视它们之间的相互依赖性。在国际化过程的初始阶段，由于缺乏信息，以及对将来的发展不能进行十分准确的预测，大部分的决策是在不确定的条件下做出的。在最后阶段，计划的重要性降低了，更需要实用主义和渐进主义，才能保证打入市场能够圆满成功。

有用的国家评价指标

不同国家和地区之间的物流服务在服务质量和绩效上都有很大的差异。例如在哈萨克斯坦，出口一个20英尺的集装箱货物需要93天，在马里需要67天，而在瑞典仅需要6天。由于目标市场在基础设施、服务质量、成本、经济、文化、政治等各方面的差异，使得这些国家和地区在时间和成本上有很大的不同。总而言之，所有这些方面都对贸易竞争有着很大影响。

当制定出国外建厂的最初战略之后，对新市场的评估将是国际化进程的第一步〈参见第5.1章节〉。国际化带来的结果是，在制定国际化战略决策、选择在哪个国家经营、在哪里建厂等方面将越来越有挑战性。为了对一个国外市场进行初步评估并判断是否适合进入这个市场，除了诸如GDP、GDP增长、单位货币收入增长率、购买力平价、汇率这些基本的宏观经济指标之外，我们还需要考虑其它一些指标。

仅有宏观经济指标并不足够作为判断是否进入市场的依据，尤其是当这个国家的背景、历史、文化、基础设施、习俗和政治与企业所属国家有着很大区别的时候。因此，当需要从物流角度对目标市场所处的国家和地区进行分析时，许多与物流相关的因素和指标如运输成本、运输的准时性、运输和IT设备、海关和边境程序以及国外市场上整体物流竞争能力等都必须被考虑进去。

为了使物流经理能够对上述各种因素有一个快速全面的认识，许多机构将这些因素进行归纳总结，形成一些重要指数来从整体上对不同国家和地区进行比较。下面给出了用来进行国家评估的有用的参数：

全球竞争指数（全球经济论坛 - www.weforum.org）

全球经济论坛定义的全球竞争指数是对驱动生产和竞争的关键因素做了总结，并把它们分为九个方面：制度、基础设施、宏观经济、保健和初等教育、高等教育和培训、市场效率、技术准备和商业成熟度。国家的全球竞争指数越高，经济竞争越激烈。

世界经济自由度（The Fraser Institute - www.fraserinstitue.ca）

Fraser组织公布的世界经济自由度反应了一个国家政策和机构对经济自由的支持程度。自由的基础是自主选择、自愿交换、自由竞争、私有财产安全等。该指数是由38个数据点构成，从五个方面反应了自由度的高低：政府规模、法律结构和财权安全、健全货币使用权、国际贸易自由和信用、劳动力和商业制度。

就业和失业指数(World Bank - www.doingbusiness.org)

世界银行Doing Business设立的就业和失业指数反映劳动力制度的柔性。它评估了雇佣工人的难易，延长和缩短工作时间规则的严格性，工资外的其它人力支出以及裁员的难易和费用。各指标越高表明制度越刻板。这个机构的其它指标包括关于税收、合同的执行和对投资者的保护。

腐败敏感指数(Transparency International - www.transparency.org)

Transparency International设立的腐败敏感指数是基于调查得出的，它由商业人士和国家分析员对腐败等级评分，分值在0（严重腐败）到10（高度廉洁）之间

物流敏感指数(Global Facilitation Partnership for Transportation and Trade - www.gfptt.org)

对物流敏感指数的评估目前还在进行之中。它是由世界范围内的国际货运公司的管理人员所提供的对一些国家的物流环境的调查得到的。调查旨在找出一些物流敏感指数来评价发展中国家和工业化国家物流活动中的关键因素。评估所涉及的方面包括贸易简易化对经济的影响、贸易物流与发展中国家、贸易物流与实用评估方法、贸易物流策略等。

全球物流指标(World Bank - www.worldbank.org)

世界银行制定了一系列全球物流指标来评估国际贸易过程的效率。全球环境的特征，尤其是各个国家具体的特性如基础设施质量的不同，政策、程序、机构的不同，都对该指标有着直接影响。尽管最初只是希望政府根据该指标制定相应策略，以帮助企业提高物流水平，不过该指标还可以为物流人员提供指导作用。它提供的分类框架和评估标准能够帮助物流经理们从物流的角度分析国家评估时，得到一个对他们各自领域的整体概念。全球物流指标可以划分为几个种类，主要是成本、时间、复杂性和效率。这个分类标准是物流经理们对目标国家的市场进行深入分析的起始点。图15中列出了一些全球物流指标。

图15 – 国际物流指标图

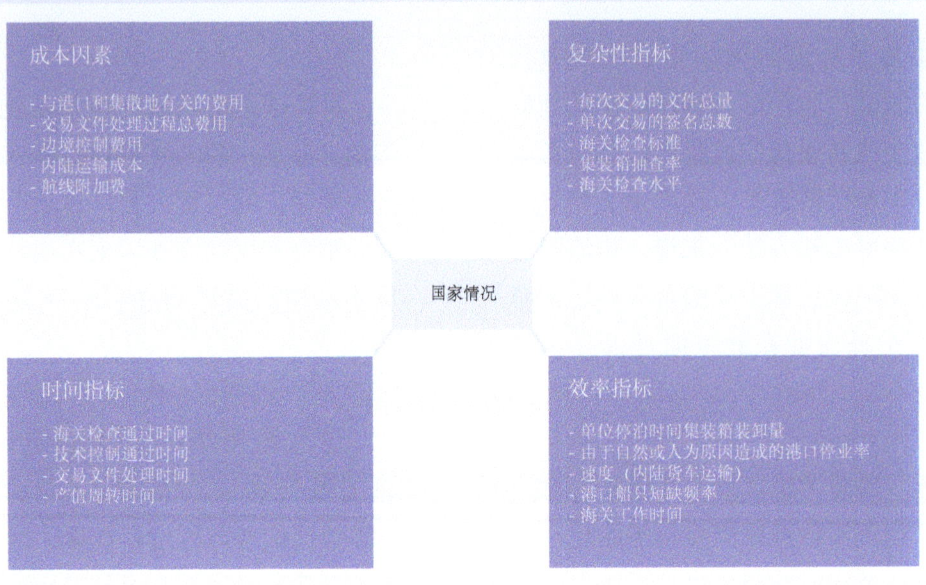

  所有这些指标都可以与宏观经济指标进行比较，并且可以以此作为标杆，从物流角度对不同国家和地区的发展水平进行比较。当在不同的国家进行经营运作时，要对不同指标分别加以考虑。此外，当对商业环境和物流服务进行评价时，也要根据实际需要将其它一些全球物流指标考虑进去。由于各国际组织、各个国家和地区实施的贸易简易化的计划各不相同，全球物流指标也会随之变化，并且要及时进行更新。这些指标能够帮物流经理们评估目标国家的现状，并为每个目标国家提供物流绩效的敏感性分析。

5.2 关于物流在国际化进程中的思考

企业为了从走向国际化的优势中获益，就必须从产生国际化想法的伊始到最终实施和集成新设施的过程中，都应考虑到物流成本和绩效水平。这产生了如下的问题：从制度上和方法上来说，物流和物流职能在国际化进程中的整合应该达到何种程度？中德企业在此方面是否存在差别？

因此，需要考虑以下两个因素：时间因素，即物流和物流职能参与国际化进程的时间；以及它们的整体参与程度。

早期物流集成的重要性。在进程的早期（如评估和制定计划的阶段），就重视物流和整合物流职能，就能确保在进入新市场时获得成功<参见 5.1章>。意料之外的物流和运输高成本常常抵消了在制造、采购、营销或销售方面国际化的优势，而且危及整个国外投资。在定义市场进入战略和设计新区域各自商业模式的阶段，就必须综合考虑未来物流系统可能面临的机会和劣势了。

对成功的项目进行观察后，我们发现：对物流情况的全面分析和对多个物流系统的备选方案进行可行性研究，不仅仅可以改善最终物流方案的成本结构，而且还可能从出色的客户服务中获得重要的竞争优势。

超过三分之二的调查企业表示，物流较晚参与国际化进程会不可避免地导致较高的物流成本<参见图16>。超过一半的调查企业认为在总体上这甚至是市场进入的潜在危机。

尽管这是参与调查物流经理的共识，但约三分之一的人感到自己没有及早参与到决策制定的过程中去，这表明并非所有企业都会利用物流知识来实现全球网络中成本结构的高透明度，从而创造竞争优势。

进入新市场的决策制定过程中物流的参与程度。在制定决策的过程中，物流经理的参与程度存在很大区别，而且中德企业也有着明显不同。60%的中国企业表示，在对是否进入一个新市场进行决策时，物流经理都会参与进来，而他们的德国同行仅有不足30%的参与度。

调查中包括三个具体的国际化问题，物流经理是否会参与到以下决策中去：（1）选择配送中心和其它物流设施，（2）选择制造设施，（3）选择供应商。大多数被调查企业的物流经理会参与物流设施的选址，但德国企业的物流经理在另两方面还是考虑得很少。而超过三分之二的中国企业表示他们的物流经理会参与选择制造设施和供应商。但是，超过25%的被调查企业表示，物流经理对最终的决策制定没有影响力。正如前面提到的，在价值链的合作伙伴中使用总成本分析以获得物流专业知识，加强成本透明度，并没有得到广泛应用。

5. 进入一个新市场的物流策略

图16 - 国际项目中物流经理的参与

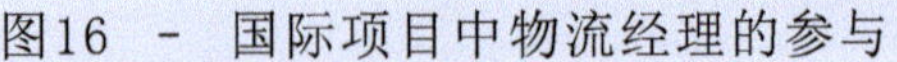

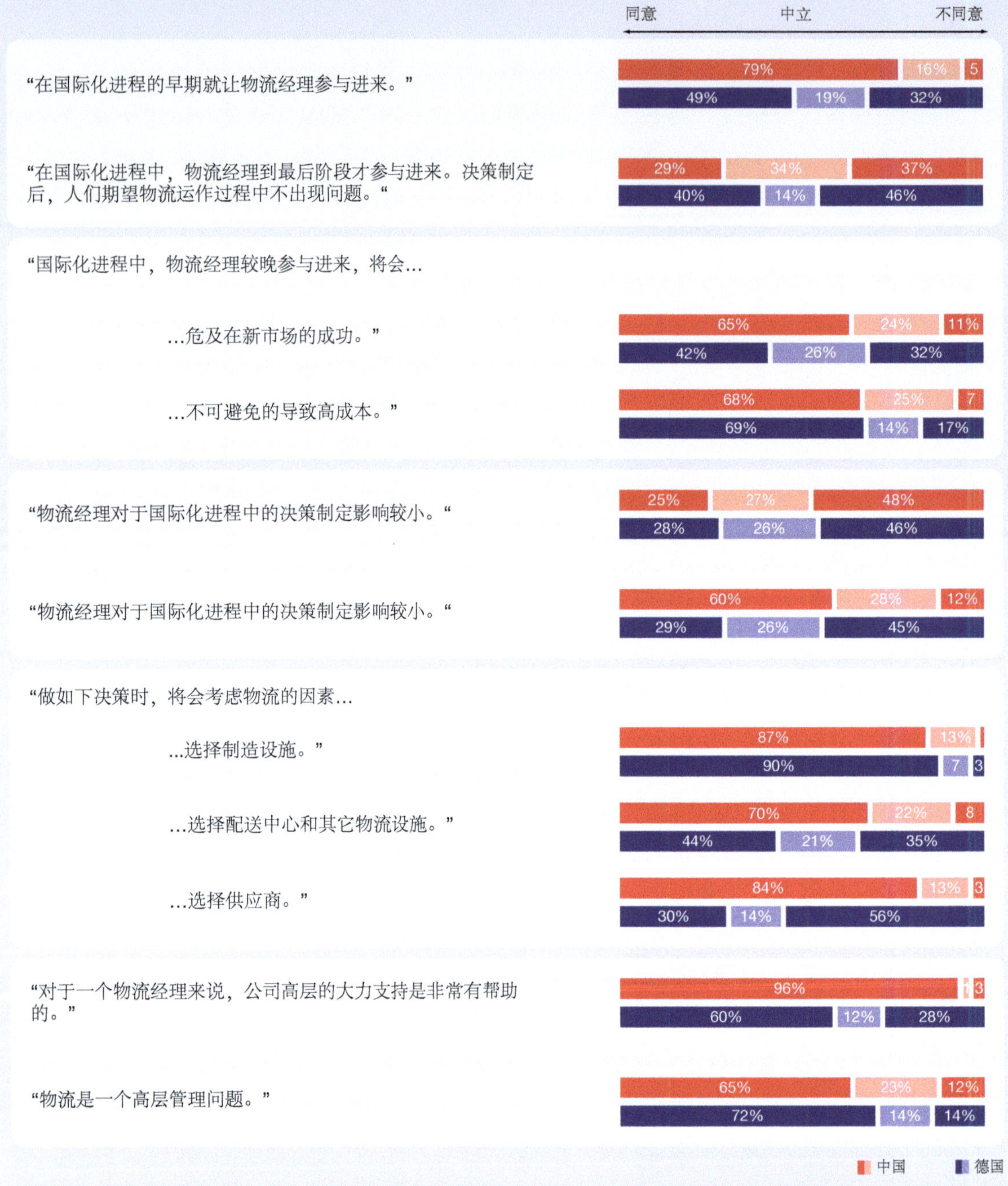

这表明许多企业还没有将物流作为客户和网络管理的一个整体和集成的工具，来发挥它的全部潜能。因此，在许多情况下，物流对国际化业务成功的战略潜力并没有发挥作用。这就是为什么超过70%的被调查企业希望得到高层管理人员更多的重视和支持，以避免失误，防止出现后续成本以及不完善的国际物流网络。

南美一览

南美洲包括12个国家：阿根廷、玻利维亚、巴西、智利、哥伦比亚、厄瓜多尔、圭亚那、巴拉圭、秘鲁、苏里兰、乌拉圭和委内瑞拉，总人口三亿七千三百万。由于竞争力水平的低下和较低的需求满足率，南美洲现在被认为是继亚洲之后，进行国际化最有利的地区。南美洲人口总数占世界总人口的5.8 %，国民生产总值占全球GDP的4 %。

从二十世纪80年代开始，南美洲各议会中就开始了民主化进程。不断提高的民主化程度使得南美洲拥有更稳定的长期投资环境。同时，除了一些特别的个例，南美已经走过了国有企业垄断的工业时期，引进外资以及建立合资企业能够有效地加速国家的经济发展。当然应该看到，现在仍然还有一些由社会主义者领导的政府反对实行自由的市场经济。

南美洲经济不稳定的一个典型例证就是上个世纪末发生的阿根廷经济危机。在那次危机中，阿根廷经历了一场由高负债利率和过高估计比索对美元汇率而引起的经济大滑坡。经济危机所带来的后果是，失业率高达23%，而货币则贬值了30%，贫困线以下人口比率达到57%。由于国际市场具有高度的关联性，阿根廷的经济危机也波及到了巴西和委内瑞拉等国。至今为止，南美洲还没有完全从这场危机中恢复过来。

在南美洲还有许多亟待解决的问题。南美洲市场是一个进出口市场。2005年，在拉丁美洲的所有出口贸易总额中，洲内贸易仅占到11%，而与此形成鲜明对照的是，欧盟内部贸易总额占到了欧盟出口贸易总额的三分之二。而即便是这少得可怜的南美洲内部贸易也只是局限在几个有限的大型城市之间。购买力高度集中在城市范围内。实际上，有四分之一的南美洲人口居住在洲内排名前十位的大城市中，多达80%的拉丁人口都是低收入者。南美洲的失业率超过了10%，而通货膨胀率则是美国的两倍，贫富差距正在不断加大。

由于受到复杂的关税条例影响，大多数南美洲国家的通关过程繁复缓慢，通关成本也较高，这就无形中增加了物流服务的困难性。税费和获取供应执照增加了投资者的成本，并因此成为了阻碍私人投资的主要因素。这些税费相当之高，以委内瑞拉为例，税率达到了利润的90%甚至更高。因此，对于投资者来说，每个国家赋税系统的设计和结构都将影响到投资者所面临的金融风险。利息率的波动，通货膨胀以及政策的不稳定也都是需要投资者慎重考虑的内容。

尽管可能存在众多的经济和政治风险，12个南美洲国家还是在2004年达成共识，建立了与欧盟共和体类似的南美国家联合体。而且在此之前，南美洲内已经建立了两个自由贸易组织，一个是在1969年成立的安第斯山国家共同体，另一个则是成立于1991年的Mercosur组织。其目的都是为了推动自由贸易的发展和便于商品、人员和资金的有效流动。这些组织的建立有效地拆除了各国间的贸易壁垒，加强了各国的贸易往来。另外，相关的法律条文也相继出台以促进洲内和洲际贸易。所有这些都预示着一个更加开放的市场和更大规模的贸易量。

从物流的角度来看，面临的最主要挑战是基础设施的建设。现代化基建和稳定可靠的通讯网络的缺乏造成了物流水平的低下以及物流成本的高居不下。根据一般的标准，物流成本应占产品总成本的6%-10%，而基础设施缺失情况下的物流成本则高出标准许多，占到了产品总成本的12%-20%。大多数的国际货物都采取航空运输的方式，其余的则采用海运的方式进行运输。即便如此，仍然没有一家南美航空公司的货物周转量能够跻身于全球前30名之列。绝大多数的全球知名物流服务提供商都以其北美总部（比如美国的迈阿密）为中心基地，来为南美洲的客户提供服务。由于地理和地形的原因，货物很少通过铁路来进行运输。亚马逊河，安第斯山和热带雨林这些特殊的地理地形都是铁路运输中难以逾越的障碍。同时，由于不同国家间的铁路规格并不相同，因此这些铁路的交界处也非常不适合进行货物的装卸。

一份最近的来自乔治亚州理工大学的研究表明，2004年，67%的南美企业使用了第三方物流服务商提供的服务，2005年，这一比例上升至72%。其中约有77%的企业与第三方物流服务商的合作关系非常成功。物流服务提供商认为南美洲市场是一个充满挑战的市场，因为他们不得不应对政治和经济的不稳定，港口和交通基建的缺乏以及赋税的问题。不断增长的物流市场中的主力是国外投资者，同时国际物流的需求也以每年10%-20%的比率增长。但也应该看到，虽然南美洲的物流市场不断发展，但是同时也存在很多问题需要解决。比如其中一个问题就是在国内工业企业的物流运作过程中，缺乏补货和运输的基本物流概念。

现今主要是外资的增加推动了南美洲的经济发展，因为这将迫使国内企业参与竞争并降低成本，同时将南美洲带入全球的竞争市场。

5.3 国际化流程的构建和标准化

对于企业而言，国际化（在外国设厂办公）并不陌生。对于不同的国际目标市场，企业会经常采用国际化方案。因此，在经过分析之后，从特定的程度上，这些国际化的项目和相关流程都会由物流人员统一处理，除非他们对每一个不同的国际市场进行单独处理。数据表明，有些物流人员会将各个不同的市场当作一个独立的项目来处理，从而在这些过程中没有什么标准化的流程可言。

接下来，我们分析了当企业在国外市场设立分支机构时是否使用了物流策略，以及物流策略的目标是什么。

国际化流程的标准化。看了第5.1章的国际化流程模型，我们相信，无论在哪个国家，任何一个项目都会包含此模型中所包含的流程。市场评估、可行性分析、对国家物流设施的评估、对设备地点的考察、物流网络的设立以及其他相关的活动，这些都是可以标准化的，从而能够提高这些流程的质量和可靠性。

标准化具有相当大的作用：超过70%的被调查者认为标准化能够提高质量、速度、以及降低为国际化所付出的成本＜参见图17＞。

尽管已经达成了共识，但并不是所有的企业都能够有效地实施上述举措：超过45%的中国企业和34%的德国企业认为他们的国际化流程并没有得到充分的标准化。因此，大多数被采访者都呼吁希望在未来有更多的标准化流程＜参见图17＞。

标准化尽管很有价值，但是也有局限性。由于在不同的国家和地区间存在着很大的不同，因此也必须对各个不同的市场进行单独处理。而标准化的流程只是进入新市场时的一个大致的框架，并没有提供太多细节分析。通过对那些成功的企业进行评估，我们发现他们都有着较高的标准化，并从中获益＜参见图13＞。但是，依然有17%的中国企业和37%的德国企业由于不同国家之间差异性而忽略了标准化的实施。

那些有着标准化流程的企业通常都有通用的指导手册，提供对于市场进行评估的方法，甚至会成立一个专门的专家小组来支持企业国际化的发展＜参见图18＞。在这样的团队中，超过半数的团队拥有物流经理。无论是否已经实施标准化，所有公司都应该从以前的市场准入中学到经验并为日后的新项目做好准备。

适应标准化需要一个过程，国际化进程中的所有决策并非都能够基于之前的分析。在调查中，44%的中国企业和20%的德国企业认为，很多时候经理的个人决断有着重要影响。战略计划具有一定的局限性，在国际化进程的后期更是如此。

总而言之，数据显示德国公司比中国公司能提供更高的物流标准，此外，德国企业还对未来的标准化有着更高的期望值。这归因于德国企业有着更多的国际化经历，以及实现了更多的国际市场准入。

为国际化所实施的物流战略。物流的发展使其具有了更多的功能

图17 – 国际化进程高标准化的现况和益处

〈参见3.1章〉，并使它在企业的成功中扮演了更重要的角色，于是，物流战略作为企业战略的一个子部分也就应运而生了。除了研究国际化流程的标准化之外，本调查还对企业在进行国际化时是否使用了物流战略进行了调研。

近一半的中国企业和大约三分之二的德国企业为国际化制定了物流战略，包括了战略目标、长期计划和方针。大多数的物流经理都在战略规划阶段得到了更高层领导（例如最高管理层）的指示或工作目标〈参见5.1章〉。在很多情况下，企业期望物流系统的性能与在国内市场中有着同样表现，并且成本也能控制在与国内相当的水平〈参见图19〉。

那些在进入国外市场时没有物流战略或没有制定物流战略计划的企业，可以用"渐进主义"或"蒙混过关"来形容。

由于全球环境复杂而且变化迅速，因此战略计划不可能涵盖其中的所有方面。为了在这样的环境中运作，企业进入国外市场时需要遵循一些步骤，从而帮助企业快速适应不可预见的变化，发生问题后能够立即做出处理。鉴于国际化过程中会面对极高的不确定性和不可预见的变化，循序渐进不失为某些企业的不错选择。

对比德中企业后，我们发现，在企业物流系统的国际化过程中，中国企业更青睐采用循序渐进策略。另一方面，很多企业在进入新的市场时采用Greenfield的方式，而且大部分的成功者〈参见图13〉甚至应用很复杂的物流计划。因此，很难归纳出一个明确的趋势。

在国外开展业务时的物流目标。当进入一个新的市场时，企业可能有着多种不同的物流目标。在我们的调查中，有

表3 – 进入新市场时的主要物流目标

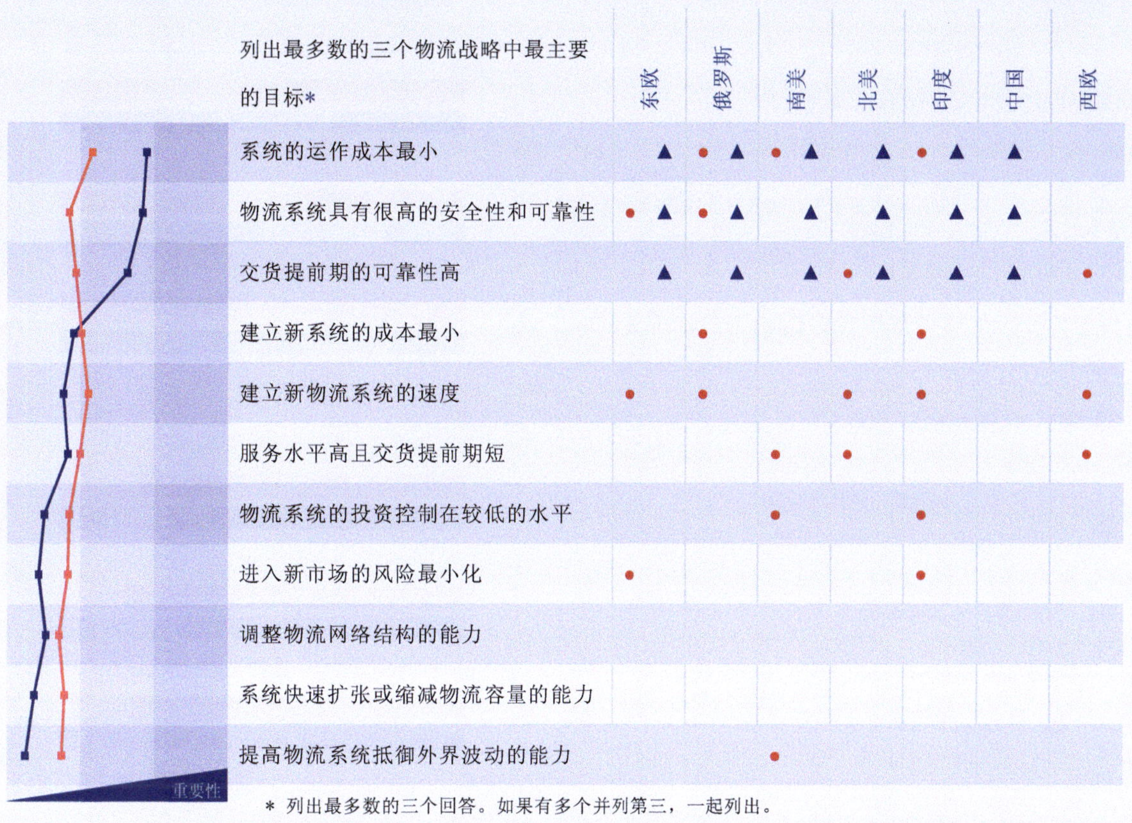

列出最多数的三个物流战略中最主要的目标*	东欧	俄罗斯	南美	北美	印度	中国	西欧
系统的运作成本最小	▲	●▲	●▲	▲	●▲	▲	
物流系统具有很高的安全性和可靠性	●▲	●▲	▲	▲	▲	▲	
交货提前期的可靠性高	▲	▲	▲	●▲	▲	▲	●
建立新系统的成本最小		●			●		
建立新物流系统的速度	●	●		●	●		●
服务水平高且交货提前期短			●	●			●
物流系统的投资控制在较低的水平			●		●		
进入新市场的风险最小化	●				●		
调整物流网络结构的能力							
系统快速扩张或缩减物流容量的能力							
提高物流系统抵御外界波动的能力			●				

* 列出最多数的三个回答。如果有多个并列第三，一起列出。

● 中国　▲ 德国

图18 – 国际化进程的标准化

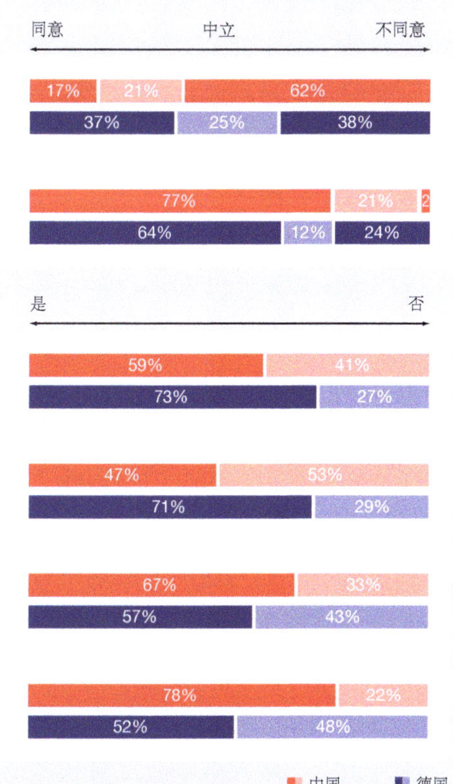

"就像每个国家各不相同一样，每个项目也各有特点。标准化程序对国际化进程来说是没有意义的。"

"在国际化进程中，更多使用标准化是有利的。"

"我们公司在国际化进程中有一套指导方针和手册来遵循。"

"我们用标准化方法和列表清单来评估新市场。"

"我们有一个特别团队/任务小组参与国际化进程，以其专业知识和经验来为这些项目提供支持。"

"物流经理是该"国际化"团队中的一员。"

图19 – 进入新市场的物流战略

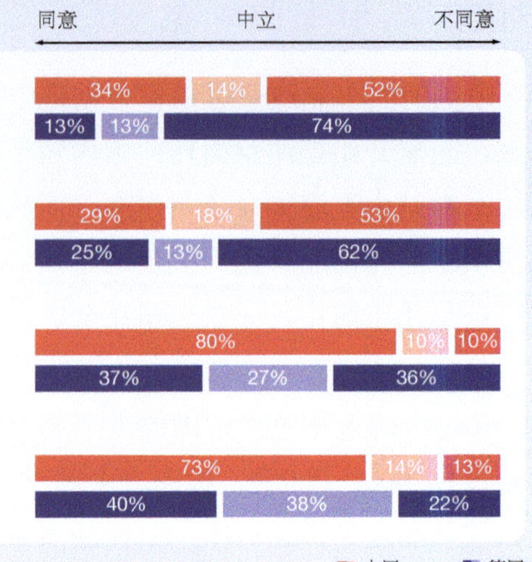

两种截然不同的目标。第一种，目标直接与国际化过程本身相关，例如，物流系统投入使用的速度、最小化建设新系统的成本等等。第二种，目标与最终系统的特性相关。例如，合理的运作成本、配送时间的高可靠性等等。

有趣的是，对于不同的地区而言，德国公司认为最重要的三个目标是一样的，它们都把注意力集中在最终系统的质量上：最终系统运作成本的最小化、运输提前期的高可靠性、运作过程的高稳定性〈参见表3〉。因此，国际化过程的结果要比国际化过程本身的速度和成本要重要得多。最关键的是最终的物流系统。

相比之下，中国公司的目标更多样化，而且根据地区的不同而存在差异。对中国公司的调查表明，建立并开展业务的速度，是中国公司最重视的物流战略问题。随后是最终系统的运作成本、建立新系统成本的最小化，以及在较短提前期下提供高水平的服务。图14中的结果显示，中国企业注重开设工厂以及开展业务的速度和成本，另外，中国企业国际化的过程的持续时间要明显地比德国企业短得多。

总而言之，很难明确地界定出中国受调查者的目标。对于不同的地区，制定一些不同的目标是最重要的。但是，从总体上看，每个不同目标的重要性似乎又是差不多的。一个可能的解释是，德国企业在任何地区都只注重有限的物流战略目标，相比而言，中国物流人员的"蒙混过关"是更高层次的"蒙混过关"。

中国的汽车工业一览

中国的汽车制造业可以作为见证中国经济快速发展的一个典范。自从70年代改革开放以来，国家开始把注意力转向国内轿车和卡车的生产。到2000年为止，上海市几乎所有的乘用车都是由上海大众公司生产，而武汉的乘用车市场基本上也都由东风雪铁龙公司的产品所垄断。同样北京也大批量地采购低成本的天津大发轿车来组建首都出租车队。总的说来，截至几年前止，大众公司在全国轿车市场的占有率达到60%。

在过去的五年里，中国的汽车制造和销售几乎增长了三倍，2003年，随着全国汽车总产量超过4，000，000辆，标志着中国已经与汽车生产大国德国和法国一起跻身于世界汽车制造国前五名之列。2007年，中国的汽车年预期销售量为49，000，000辆。到目前为止，虽然外国企业在中国汽车行业的急剧扩张中扮演着很重要的角色，但是主要是通过中外合资的形式实现的。

国内外汽车生产商在过去两年里的大量投资导致了汽车生产过剩的问题，而由于尼桑、福特、丰田、现代等国际知名汽车制造商也进入到中国市场，大众、通用、本田、雪铁龙和其它汽车制造企业正面临着生产空间变得日益拥挤的局面。2003年，汽车生产的增长比率为87%。而在2001年到2005年间，汽车零部件的进口增长比率在17.4%到25.8%之间。由于中国进口的所谓"国外原装"的汽车受到严格的关税限制，现在在新增加的汽车中只有5%是"国外原装"，与此同时，大众桑塔纳2000车型的零部件已有近98%是由中国本土生产的。这里将列举出一些国外汽车生产商的实际数据，使我们能对中国快速增长的汽车行业有进一步的加深了解。

- 德国大众汽车公司2003年在长春投资了5，300，000，000欧元，并且在上海投资6，000，000，000欧元建成了一个年产量达300，000台的发动机制造厂。
- 2003年美国福特公司在中国的投资为1，000，000，000美元，并且计划在2010年前，再追加16，000，000，000美元的投资。
- Delphi公司计划在中国进行喷射泵的开发工作。
- 雅马哈公司打算将其研发设计部门迁至中国。

5.4 如何应对动态的环境

第五章的第一部分主要论述国际化流程，如它是如何组织的，由哪些步骤组成，如何把物流整合到决策中，以及公司遵循哪些目标。第二部分则着重指出了在建立国际化的物流系统时必须考虑的问题。

用于设立境外企业实体的物流战略的关键要素。无论一家公司是否从事战略规划或相关活动，根据渐进主义的观点，不管物流策略试图达到什么目标，属于任何物流方法都必须包括三个关键的要素，以确保最后的物流系统能有效地支持外商进入市场：（1）物流系统必须能够应付运作中瞬息万变的情形；（2）全球物流标准化进程必须与本地的情况进行权衡〈参见5.5章〉；（3）做出决定时必须考虑与物流服务供应商的战略合作〈参见5.6章〉。

瞬息万变的商业环境。为了将物流作为企业竞争优势的来源并充分获益，管理者制定并实施了那些越来越复杂、持久、难以扭转，且风险比以往都大的物流系统。具有讽刺的是，最大的风险在于不能成功建立可以对迅速变化的环境做出反应的物流网络。正如第4章指出的那样，国际化在很大程度上给物流网络增加了动态性和不确定性。物流条件如基础设施、法律法规以及行政事项等，都是在不断变化的。市场未来的发展，甚至公司未来的成功，从本质上说都是难以预测的。鉴于上述这些因素在每一个国家有着不同的变化，显而易见的是，设计和管理全球物流系统具有极高的挑战性且容易出错。相应地，超过80%的物流管理者强调，商业环

境的动态变化为国际化创造了最大的挑战〈参见图20〉。

在未来，物流系统需要应付难以预测的和迅速变化的要求。这一点无可否认，从而导致了一个问题，那就是物流系统在规划和设计阶段的灵活性和敏捷性应该达到何种程度〈参见图20〉。

考虑动态性和不确定性。被调查的超过80%的公司坚持认为，AS-IS的情形不能够为规划物流网络提供充分的依据。在规划和设计物流能力、设施的位置、网络结构、提前期的定义等的过程中，基于企业对自身未来的期望，80%的中国企业、60%的德国企业物流发展情况是不同的。在与公司的其他部门合作时要考虑到这一点，同时还要结合其他专家的期望，如：销售期望。

但是，这只是对将来发展的估计，很可能会很不准确。因此，72%的中国和53%的德国受访者侧重于创造灵活的物流结构来应付不确定性。

多种原因都能导致动态性及不确定性，例如：顾客的偏好随时间变化而改变、法律限制的改变等等。在中国的经济回升的早期阶段，举例来说，外国投资者只在沿海地区独家设立生产设施和分销中心。然而时至今日，中国的购买力正越来越向腹地延伸，侧重于沿海地区的物流结构未必会仍然有效。

若谈及在法律范围内的改变，人们可能会想到中国入世。对于那些适应改变的

图20 - 国际物流中的动态特性

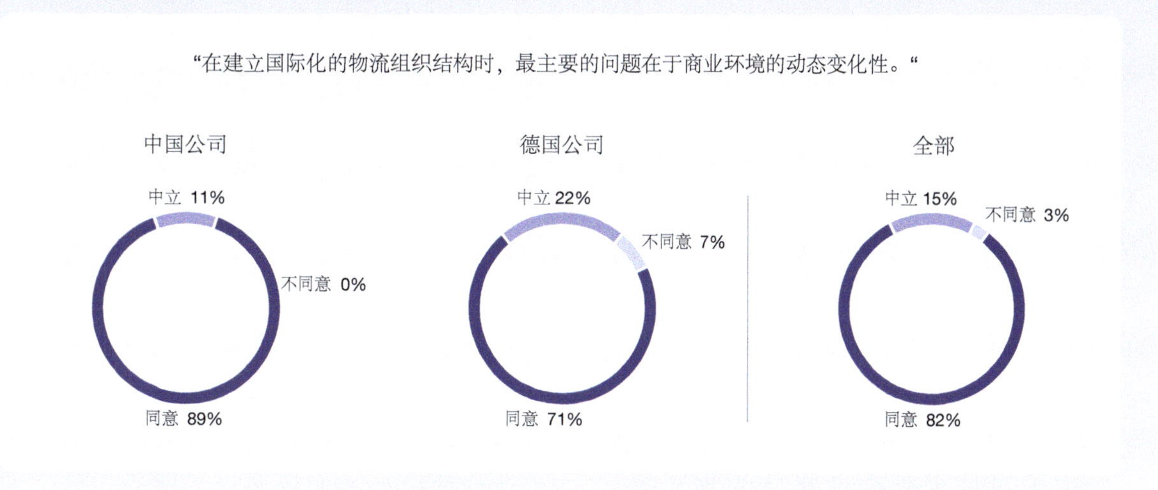

企业而言，中国入市创造了许多的机会，而那些没有立即意识到转型会给企业带来怎样后果的企业，可能会很快退出市场。

一般说来，在大多数新兴的市场中，最容易出现动态和不确定的环境。然而，动态性即使在发达地区也不容低估，一个突出的例子就是欧盟的东扩、当一些国家如波兰、斯洛伐克和捷克共和国加入欧盟时，许多公司认为必须对它们在西欧的集中分布结构进行重新设计，许多中央仓库被拆除东移，以配合扩大后的市场。

应对动态性及不确定性。物流适应性有着不同的层面，从系统适应顾客短期需求变化的能力，到改变网络结构的可能性各不相同，这主要取决于其地理中心的变化。只要符合公司的经营模式，各种方法都可以应用，例如延迟策略、协议弹性工时、外包观念等等。十多年以来，物流外包就是一个热门话题，当进入国外市场时，物流外包就变得更加相关了〈详见5.6章〉。

比确定安排以提高灵活性更为重要的，是管理者在战略层面的全局意识，即他们在国外市场建立的物流系统必须足够敏捷，以满足将来难以预测的业务需求。

防止撤退出市场的策略。当进行国际化商业扩张时，管理者往往会提供充足的事例论据，来证明能够获得国际成功。但同时也有许多著名的失败案例，如公司失去了海外财富，甚至决定退出某些市场。沃尔玛从德国零售市场的退出是最近的一个典型例子，另一个则是德国家用建材连锁店OBI，他们几年前决定撤出中国市场。

市场发展的高度不确定性，对市场的发展形成了挑战。物流系统应当为迅速扩张提供支持，同时也需要考虑到退出某个国家的可能性，为退出提供最低的沉没成本和剩余间接费用。

即使被调查公司大多都反映，在规划和设计国际物流系统时，他们认为系统的环境以及条件可能改变得很快，但仍然没有很好地考虑到在新市场中会遭遇失败的情况。

企业普遍对国际化项目的成功具有非常高的期望，因此，有三分之一的受访者把任何国外市场进入作为一项长期的契

图21 – 物流战略中的退出之考虑

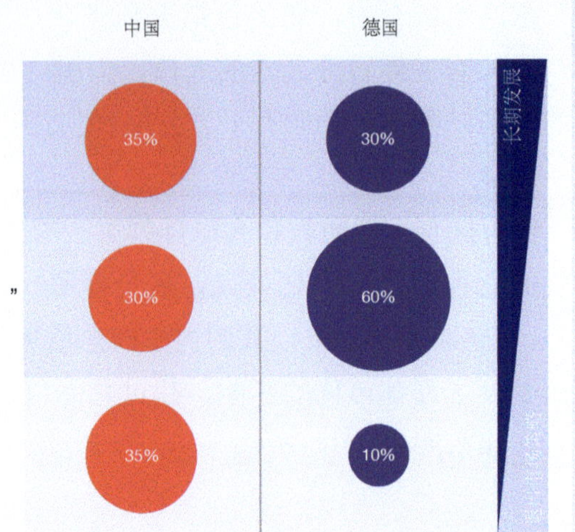

约〈参见图21〉。当意识到市场风险很大时，60%的德国物流学者会在他们的策略中包括一个出境选择权，但是只有30%的中国同行有同样的行为。另外35%的中国企业，以及只有10%的德国企业考虑到可能在国外市场遭遇失败，且在任何新的物流体系中连续使用出境选择权。在把灵活性体现到企业的国际化这一点上，中国公司似乎比德国公司做得好，这可以解释为中国公司比德国公司面临着一个更为动态的环境这样一个事实。

从物流的角度来看，物流环境的动态性及不确定性难以被影响。因此，企业必须处理好这种环境。国外市场进入目标中灵活性的高低，仍有改进的余地。结果表明，与定量的方法相比较，例如成本和性能，灵活性则相对得到较少的关注。鉴于当今商业环境的动态性和不确定性，以及僵化的物流体系中相应的成本，这个目标体系似乎值得怀疑。

北美一览

　　在这项调查中，北美包括美利坚合众国（美国）和加拿大。北美作为排名第一的经济大国，它只有世界百分之六的人口，但国内生产总值却占了全球33%的份额。北美是世界第三大的大陆，面积约1850万平方公里（仅美国和加拿大）。

　　作为拥有土地面积和人口排列第三的大国，　美国是世界上经济力量最强的国家。其国内生产总值超过13万亿美元，居全球第一。美国还是第一大进口国和第二大出口货物国。此外，　自2000年以来，其国内生产总值每年增长2.8%，　2006年失业率已下降到仅4.6%。美国还受到了高都市化的影响，超过总人口（3.02亿）的87%的人居住在城市地区，主要集中于西海岸和东海岸。

　　美国拥有便捷的州际道路系统和高速公路，以高效率的道路网络连接着几乎每一个城市。而另一方面，铁路系统则相对稀少。全国铁路客运公司——美铁，为很多城市提供服务，但在有些地区，特别是贯穿东西部的地区，铁路运输仍然不发达。例如，为了横过东西大陆，有时需要花费一个星期，鉴于此，很多公司宁愿使用较为昂贵的空运。

　　因此，　美国的航空业从一开始就显得格外重要。距离太远，使得空中运输得到了广泛的发展。美国拥有全世界最高比例的乘客居民，据报道，2005年有6.609亿名乘客，意味着每一个居民每年至少搭乘两次飞机。美国有7个机场属于世界排名前15的货物装卸机场，同时恰好有8个机场拥有前15位的客运能力。这两份名单都是由美国机场与孟菲斯和亚特兰大提供的，孟菲斯是货运老大、物流服务提供商联邦快递的总部，亚特兰大则是排名第一的客运机场。联邦快递（FedEx）、联合包裹服务公司（UPS）和联合航空公司的吨公里收入（RTM）是世界领先的。航空类公司，如德尔塔航空公司、美国航空公司和西南航空公司则是旅客运输行业的排头兵。随着2005年新的航空协议接近尾声，美国与加拿大达成了全面开放航空运输的协定。该协定对两国之间运营的航班数量不加限制，同时也取消了各种飞行费用。

　　美国主要的海港在东北部，发送来自欧洲和沿整个西海岸的货运。美国三个最重要的海港位于洛杉矶、长滩和纽约，负责美国近五成的集装箱装卸。从英属哥伦比亚到美国南加州的西岸港口之间有着激烈的竞争。北美的西海岸是一个世界级的物流中心，为进口亚洲货物提供了便利，特别是在中国开始崛起之后。中国每年向美国成功运送超过1200万个标准箱，这些为港口带来了困难，如土地限制和基础设施达到或接近最大生产能力等问题。

　　加拿人的官方语言是英语和法语，它是北美地区占地面积最大的国家，也是世界上第二大的国家，仅次于俄罗斯，人口密度为每平方公里3.5人（总共3300万），是世界上最低的。许多农村地区甚至无人居住。加拿大属于富有的国家之一，每人平均拥有三万美元以上。加拿大的经济在过去几十年中一直稳步增长，失业率较低，约为6%。与其他发达国家，例如西欧相比，初级产业在加拿大有着非同寻常的重要作用。石油和伐木业是加拿大最大的

行业。加拿大的石油和天然气储量居世界第二，仅次于沙特阿拉伯，这使它成为了一个能源出口大国。加拿大同时也是世界上主要的自然资源生产国，如镍、铝、锌、铀和黄金。然而，服务业掌握着加拿大的经济命脉，从事服务业的人数超过了75%。加拿大高度依赖于国际贸易，尤其依赖与之接壤的美国，从而外部的不稳定性直接影响到本国经济。

由于国家的疆土辽阔，公路或铁路运输难以涵盖各个领域。然而，加拿大拥有世界上最发达的基础设施。加拿大政府在2007年作出了一项具有历史意义的决定，为今后7年的基础设施建设投资了160亿美元，这预示着未来的迅猛发展。加拿大兼顾了欧美的优势，居住密度与美国类似，而运输则具有欧洲的特点。

尽管针对1994年1月签订的北美自由贸易协议(北美)的某些部分仍有争论，但人们普遍认为，这项协议对三个签署国——美国，加拿大和墨西哥都非常有利。北美自由贸易协议生效已经有13年了。当1994成立年北美自由贸易区时，它代表了6万亿美元的国民经济总值与3.6亿多的人口。在短短十年里，这三个签署国就增加到了12.5万亿美元的国民经济总值和4.3亿多的人口。该协议是用来处理三个签署国——美国、加拿大和墨西哥之间的各种贸易壁垒和关税，取消了很多种类的限制，包括汽车、纺织品、计算机应用课程和知识产权。

自从签署北美自由贸易协议以来，美国和加拿大的贸易都出现了蓬勃增长，出现了更多的专业化以及两地经济的融合。但是，尽管该协议放宽了贸易和投资政策，它们却几乎没有加快双边贸易的自由流通，也并未增加为了发展双边经贸所需的流程和文件。这些流程阻碍了运输的自由流通，由此产生的费用成为了另一种形式的非关税障碍，导致了物流成本的增加。美国和加拿大拥有世界上最大的贸易关系，2006年两国之间的商品贸易总额为5337亿美元。美国依然是加拿大最重要的市场，加拿大是美国最大的贸易伙伴。2006年美国从加拿大进口3034亿美元的货物，同时美国向加拿大出口货物2303亿美元。美国和加拿大同时也是对方主要的外国直接投资的来源。

第三方物流业在北美两个最重要的变化，是全球化和保护业务连续性。全球化的主要趋势之一是，越来越多的美国和加拿大公司在产品装入出口集装箱之前，在海外出发点对产品加以控制。这一趋势在零售业中表现地尤为明显。在业务连续性方面，顾客希望他们的第三方服务提供商是一个真正能够确保其业务持续不间断的全球服务商。顾客通过加快频率、促使供应商快速反应、促使供应商在不影响正常业务活动的情况下交付货物等方法来调整自身的采购计划。在过去的十年中，制造生产逐渐转向了低成本国家，尤其是中国。例如，美国从亚洲进口产品的70%几乎都来自中国。

5.5 全球标准和地方适应性

除了动态性的不断提高，内在的异质性过程是对全球物流网络另一项重大的挑战。这主要是由于涉及大量地方和中间商，以及在不同的物流条件下，用不同的方法和流程与供应商和客户工作的情况，使得全球物流管理成为一项日益复杂的工作<详见4章>。

标准化进程被认为是可以解决复杂的全球物流网络的方法，同时也得到了广泛认可。与第5.3章中描述的、试图对国际化进程中重复的流程加以标准化这一概念相类似，全球物流网络中的物流活动可能也需要进行标准化。这意味着复杂的物流活动将被分解为数量有限的简单标准化过程模块，这种模块在全世界所有地方都可以用同样的方法实现。此外，完整的物流实践同样适用于网络中的任何地方，无论发生在公司的国内市场，还是海外市场。

全球标准化进程的潜能和特色。全球进程的标准化可以在许多方面减少问题的复杂性。首先，它使企业可以利用与全球相连的以及统一的信息化系统来监控他们的网络，以建立全球化的透明度。如果所有过程都有最少相似性，则一个预先设定的KPI能涵盖全球供应链，能使各地的网络具有可比性，包括质量、时间和成本。在遇到破坏或阻止的情况下，更容易发现和解决有关问题，从而促进全球范围内的流程最优以及最佳的执行情况。此外，当更换工作人员和管理者对不同的设备进行管理时，可以使用尽可能少的指示。以上这些都是全球标准化进程的特点，加强了对物流网络的监督和管理，降

低了成本，节省了时间，提高了可靠性。

对于单个公司而言，由于其所在的行业和商业模式有所不同，因此各种需要进行标准化的流程会存在差异。基本上所有入库的、内部的和对外的进程都能够被标准化。标准化包括用箱子和托盘来运输和储存货物、定单循环、补给策略、应用准时制和看板系统、生产调度、推拉战略、交叉船坞以及仓库的设计和布局，分销网络在多大程度上是中央的或分散的，使用自动编号技术（如条形码和无线射频技术）等等。标准化绝不仅仅局限于一家公司，它也可能包括供应链中的合作伙伴、客户和供应商，实现协同流程和接口的规范化，从而加强合作。

全球标准化的局限性。尽管优势显著，标准化的概念还是存在局限性。不同国家和地区的本地商业环境存在差异，如法律框架、基础设施和顾客期望等，需要物流过程与当地情况相适应。

举例来说，德国零售商麦德龙对其现购自运的设计和规划有一套固定的模式，过程相当的标准化。例如：用货车进行补货时必须在出口的后端卸货，且在一天的什么时间交付哪些类别的物品都有明确规定。容易腐烂的货物有自身特殊的时段，而干燥的食物又不同，诸如此类。在众多麦德龙分店中，这些标准化的观点和流程，已被证实提高了它们的实用性和效率。

然而在不同地区，客户有不同的习惯并喜欢处于不同位置的商店。欧洲消费者

图22 －　国际标准化跟按照地方性的需求改变的平衡

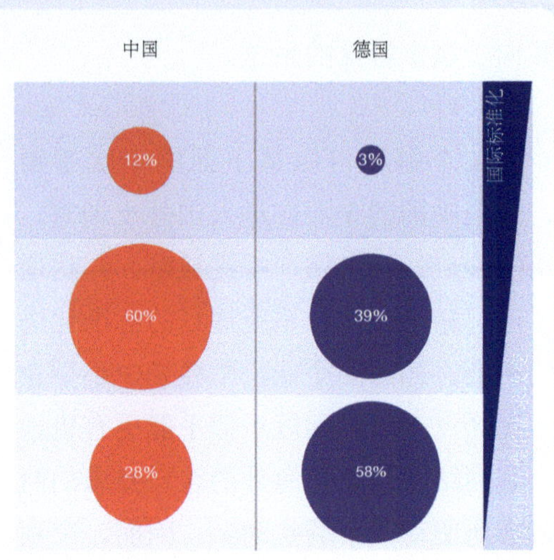

普遍喜欢位于靠近城市中心商店，而美国的消费者喜欢在郊区综合商场。中国或印度的消费者则喜欢在住宅区附近的小商店购物。不同的购物方式对商店和商品品种的类型有着不同的要求，例如，在许多发展中国家，由于收入低且缺乏贮藏和冷藏空间，人们会每日或根据需要购买物品并运送回家。因此货物需要提供小包装，且较多数量的小商店只提供特定类型的商品。

根据消费者行为的特殊性，以及亚洲房地产市场的实际特征，麦德龙对外围建筑的标准化设计进行了修改，以应付这些市场中的本地需求。这是一个典型的例子，是公司在全球化时为了适应本地情况所必须考虑的。

全球标准化与地方适应性。上面的评论和例子引出了一个问题：公司在进入国外市场时，需要在何种程度上应用标准化的物流过程。在进入国外市场时，需要如何适应以满足目标地区的需求。

如图22所阐明的，12%的中国和3%的德国受访者表示，实施物流系统时无需顾及本地适应性。这些公司只是简单地把其国内系统的设计和结构移植到国际市场，而没有根据当地需要和国家的具体情况加以修改。这可能包括运输方式、使用仓库、交叉码头，与服务提供商、供应商的协作等等。

在许多情况下，若当地的特定条件阻碍了系统的全面实施，绝大多数中国企业（60%）接受最小化程度的地方适应性，但仍坚持尽可能使用标准化物流过程的理念，约39%的德国受访者遵循相同的做法。

大约58%的德国企业为每个市场开发出全新的物流系统，以确保其尽可能地符合当地的条件，约有四分之一的中国同行采取同样的行为。

中德两方存在的差异有两种解释：首先，中国企业的竞争主要集中在日用品市场，在这里低成本是顾客购买决策的主要因素。因此，标准化对他们来说似乎是适当的。另一方面，德国企业提供的大多是高精度的复杂产品，认为自己的物流系统必须提供优质的客户服务。因此，德国企

业倾向于制定特殊的系统。

第二个解释可以从下面的事实中得出，一般中国的企业在进入国外市场时，采用Greenfield方法（推进战略），这增加了运用精心设计和系统规划的标准化进程的机会。德国的企业，主要是中小型企业，在经过一段时间以后似乎能更有机地培育其网络，由来自海外的客户要求的商业机会所驱动（市场拉动）。因而，个别的解决方案得以发展，并在以后需要时进行优化。

成功企业<详见图13>的全球化标准程度高于平均水平。然而出人意料的是，德国受访者比他们的中国同行更多地这样做。

基本上绝大多数的受访者都体会到了实施国际标准化带来的好处。每种经营模式在何种程度上需要一个怎样特殊系统，每个人都持有不同的看法，因而造成了不同的结果。除了法律和基础设施之外，可以说，总体目标是满足客户的需求，但它需要得到有效的平衡，在多大程度上认同

且重视区域适应性，以及在多大程度上能够接受公司全球化的标准。

像许多其他学科一样，"全球化思考，地方性行动"，是完全正确的，即在头脑中有一个全球的战略，并在需要的时候应用它。

5.6 与物流服务提供商的战略协作

如第5.4章所示，在国外市场进入的背景下，以下三个方面对于任何物流方式来说都相当关键：如何应付急速转变的物流网络运作环境〈详见5.4章〉；如何平衡全球物流进程标准化与本地适应性〈详见5.5章〉，以及如何与物流服务提供商进行战略协作。

物流外包。在经济迅速发展的时代，物流外包活动数十年以来一直是一个热门话题，它推动了物流服务提供商的不断发展，以及行业的繁荣。已经出现了不少类别的物流服务提供商，从货运代理和第三方物流公司到专门物流与供应链管理顾问公司。它们都提供多种物流服务，包括基本的活动，例如运输、仓储和装卸，乃至精密和复杂的网络解决方案，也包括信息处理。现今的物流收入表明了物流服务提供商行业的经济意义：仅以德国为例，2004年的物流总收入约为1700亿欧元，在被调查的西欧地区（欧盟15国，瑞士和挪威）约为7290亿欧元。在中国，物流服务（包括外包服务和自营物流活动）总值估计在2500亿欧元到3200亿欧元之间，大约占到了中国国内生产总值的五分之一。

相比较以前的产品都在本国土制造，而现在的全球化和产品制造在不同的区域，这使得物流外包活动变得更加重要。因此，制造商和零售商在国外市场设厂时，越来越依赖于物流服务提供商。

物流服务提供商对国外市场进入的重要性。在本次调查中，不到三分之一的公司表明，他们需要有自营物流资产和业务，从而为他们的商业活动维持正确

的质量，提供所需服务水平。但是大多数公司只掌控了一些少数的最重要的具体行动。制造商和零售公司极少拥有一整套的物流活动并自行管理。

当公司进入国外市场，实行全球化以减少市场进入风险，希望从合作者那儿获得应付国外事务的技巧时，它涉及的外包活动就更广泛了。很明显，现在进行物流外包活动是一个战略性决策；此外，物流服务提供商的作用已经改变，因为它们已经被允许进一步参与到客户的业务和供应链中。

在这种情况下，在新的目标市场，不能提供足够多的、且有能力的物流服务提供商，71%的中国和61%的德国管理者宁愿帮助新的物流服务提供商在市场中发展，也不愿意投资于专用自营资产。这清楚地表明了国外市场进入和全球物流外包具有很强的一致性。

因此，物流服务提供商行业需要解决这一强烈的外包意愿，为客户提供更快捷的进入市场的服务。很多企业不具备独自在国外进行物流业务的知识和经验，直到最近，在大多数情况下，这些企业的物流活动一直局限于它们所在国家或地区。因此，那些能够支持中小企业全球扩张的特殊服务有巨大的潜力。

图23 - 国际化中物流活动外包的目标

物流外包的战略目标。如前所述，时间对于成功的市场进入具有重大的战略意义〈详见5.1章〉。因此，与物流服务提供商协作的最高目标，就是推动物流服务提供商现有的全球网络的发展，确保迅速进入市场〈诸如完整的例子〈参见5.1章〉。

对于德国的物流学者来说，物流服务提供商通过扩大经营规模而获得规模经济是最重要的。

由于建立全球物流网络需要专业的区域知识，以配合当地的文化、政治和环境的特殊性，中国管理者把物流服务提供商的特殊技能和服务质量作为选择外包最有力的依据。物流服务提供商在各自的国家或地区，已经具有了极大的本地市场专长和知识，熟悉法律和行政法规，具备所需的执照等，为外资的市场进入提供了极大的便利。

外包的另一个重要的目的，是保持较低的市场进入投资，从而以较少的资产战略减少内在的风险。此外，与物流服务提供商的合作还可以提高公司物流的灵活性。一般来说，物流服务提供商比客户更能够在短期调整能力。由于它们提供服务的客户种类多，每个公司需求的变化性很可能由于在物流服务提供商处的聚合而得到平衡。此外，76%的中国和48%的德国受访者试图通过执行短期合同来提高其灵活性。

图24 – 物流活动外包战略与依靠自己的人力物力来完成的物流活动

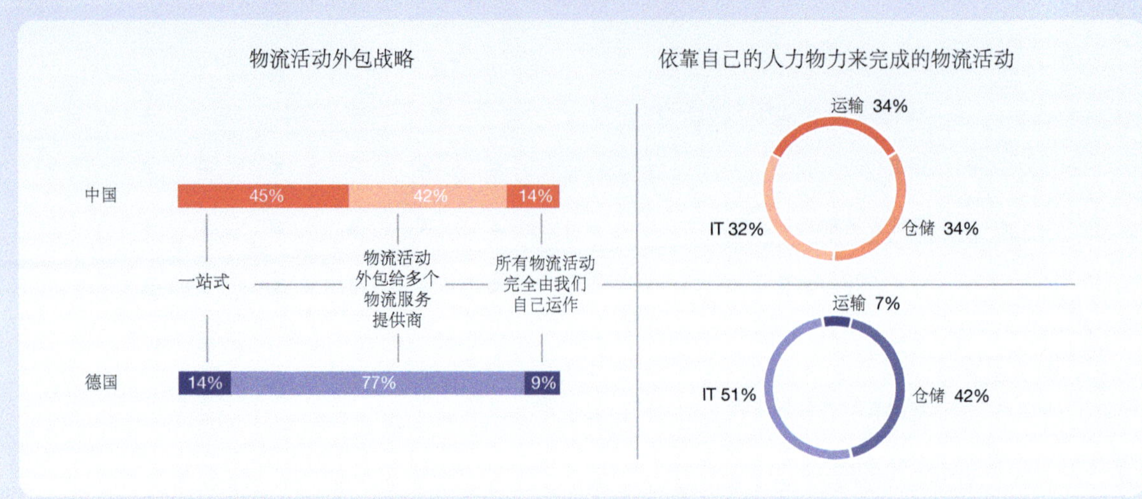

外包水平和自营物流活动。虽然中国和德国的公司都意识到物流外包带来的好处，他们与物流服务提供商具体的合作还是显示出了一些区别。一般来说，中国公司比它们的德国同行表现出大大降低的外包水平＜参见中国一览插页＞，可是那些现在正在进入国外市场，即将成为全球一员的中国公司则显示出了较高外包水平的发展趋势。

关于与物流服务提供商实际的合作，有45%的中国受访者喜欢一站式购物，即它们只与一个合伙人一起工作，这个合伙人要提供全球范围且涵盖各种所需的服务。通常服务提供商只是在地理位置上覆盖了服务的区域，然后通过与其他合伙人和分包商的密切合作而完成整体一系列的活动。但是从顾客的角度来看，优势在于只与单一的合伙人合作，从而简化了协调工作。

然而，只有14%的德国受访者需要一站式的解决方案。当客户得知物流服务提供商不能提供全部所需的服务而需要与分销商合作后，无论物流服务商是否保证所有的合伙人和分包商都积极参与，无论他们承诺提供何种程度的可靠性与质量保证，都会遭到德国客户的怀疑。此外，制造商和零售商仍避免只依赖单一供应商的模式，因为它们不希望自己的全球物流进程仅仅依赖于单一的第三方的能力。除了出现争论问题时，议价能力和成本转移是行之有效的方法，而不依靠单一的物流服务提供商。

对72%的德国和42%的中国企业来说，绝大部分的物流管理者向多个物流服务提供商外包各种特殊的业务。物流与供应链网络的整体控制仍然在客户手中，使用多个服务提供商的有限服务，能够确保企业不过分地依赖于某个物流服务提供商。

纵观各种物流外包活动，客户往往要求提供增值服务和全球化能力，但这些类别不如最终的外包决策重要。此外，当完全将供应链进行外包时，物流外包仍以传统服务为基础，并在较小程度上有更全面的方法，这主要是由客户的关注导致。客户普遍关注的是以下问题带来的潜在损失，如直接控制物流活动、外部公司服务水平的不确定性，以及雇用第三方服务商的成本问题。

表4 – 选择LSP的标准

最多数的三个选择LSP的标准*	东欧	俄罗斯	南美	北美	印度	中国	西欧
价格	▲	● ▲	● ▲	● ▲	● ▲	▲	
物流服务的质量，可靠性	● ▲	● ▲	● ▲	● ▲	● ▲	▲	●
原来合作的感受	▲	▲	▲	▲	▲	▲	
可靠性/质量				●	●		
管理者的动机与品质							●
柔性和适应性	●	●					
声誉，品牌，形象		●	●	●			●
竞争力/未来的发展潜力			●				
规模/效益	●		●				●
技术装备和技能							●
有足够的资金在未来投资，偿付能力							

重要性

*列出最多数的三个回答。如果有多个并列第三，一起列出。

● 中国　▲ 德国

在被调查对象中，只有14%的中国公司和9%的德国公司，是完全地利用它们自己的资产和设施进行物流活动的。

在国际化时，公司不完全外包的活动构成，在德国和中国的公司也显示出了很大的不同。中国受访者在运输、仓储和IT技术方面，不外包的可能性几乎相等。德国受访者则表示，自营运输是非常罕见的现象。42%的德国公司在国外市场拥有自营仓库，52%的德国受访者倾向于自行设计其全球物流网络，这个比例是最大的<参见图24>。

选择物流服务提供商的标准。当考虑应在何种程度上进行物流外包，以及选择由一个或多个合伙人处理外包业务时，就是选择物流服务提供商的时候了。

调查确定了在国外市场设店时，选择物流服务提供商的最重要的标准<参见表4>。首先，当前最关键的因素就是价格，而下一个最主要的相关因素是质量和提供可靠的服务。两者并非相互排斥，但是必须经过慎重的权衡。当公司把成本做为主要的标准时，对全面成本保持一个整体观是非常重要的。客户应该意识到，与有能力的物流服务提供商开展卓有成效的合作，可以显著地改善产品在世界各地流动的方式，从而降低成本。这种在全球物流网络中创造节约的潜力，有可能在价值上超过物流服务提供商的直接成本，因为可以通过考虑其它问题来得到提高物流效率，如质量、业务和技术等。

根据受访者的样本显示，进入国外市场选择物流服务提供商的第三个最重要的

标准是从以前的合作中获得的经验。对于中国公司而言，这更是最重要的标准。三分之一的中国和10%的德国物流学者表明，他们是不会在进入一个新市场时选择以前没有跟他们在国内合作过的物流服务提供商的。大约有五成的中国受访者始终与一个有声誉的全球物流服务提供商合作，只有三分之一的德国受访者采取同样的方式。

对不同目标地区和国家的标准加以分析后，我们得知，德国公司不会因为地区的差异而改变它们的喜好，他们前三条标准对于所有的地区来说都是一样的。相反中国的受访者就会区别对待，例如：在东欧和俄罗斯，灵活性相对有较高的重要性；在南美，他们相对更多地考虑生存能力以及他们的合作伙伴未来的发展潜力等。

德国公司在这项调查中表示出他们在选择标准上有着明显的模式；而中国公司的选择标准，是更平等地分配的，并且会随着目标地区不同而发生显著的变化，这可能是因为中国公司比他们的德国同行更习惯于循序渐进。

全球物流业。 总体而言，调查结果显示，通过其全球扩张，公司已经提高了对物流外包的期望，他们越来越需要从物流服务提供商那里获得复杂和专业化的物流解决方案，这就要求物流服务提供商不断地提高他们的能力并改善设备。因此，物流业的竞争日益激化，使物流服务合同变得更加成熟，并且给提供商带来了更大的压力，促使他们不断改进方法。

结果清楚地表明，物流服务商除了业务网络覆盖了广泛的地理范围之外，客户还特别欣赏物流服务提供商对每个地区和国家当地物流知识的了解。然而，尽管在物流市场上有着主要的全球供应商，且近年来并购活动屡见不鲜，一个拥有知识水平和基础设施，能够在未经分包与合作的情况下真正实现全球化经营的供应商，在目前却是非常罕见的。事实上，只有极少的物流服务提供商，由于它们的大小和规模，能够确保一个项目在全世界范围内得以妥善实施，使用自身的员工在所有地点直接为客户完成工作。

6 总结

根据本次调查报告所得出的研究和分析结果，能够充分地证明，物流在全球化的商业活动中起着至关重要的作用。

本次研究报告中所建立过程模型使得我们可以对企业如何进行国外市场开拓以及物流在其中所起到的作用等问题进行概念分析。来自中国和德国最成功企业的调查结果表明，将各物流要素进行紧密集成，以及物流经理尽早介入开拓国外市场的商业策划和决策过程中去，都是它们成功的关键因素。

虽然很难制定出一个适用于所有企业的标准，能够确保全球化进程的成功实施，但是面对不同的商业模式、战略目标以及不同的市场特征，每个企业都必须努力寻找出最能满足其个性化需求的物流理念。一般来说，德国公司多采用正式的科学化方法进行集中型战略计划。而中国公司则往往采用感性的主观性方法进行战略规划。但是通过对绝大多数成功的问卷回馈结果进行整理分析，可以看到一个不同的景况，即那些最成功的中国企业都具备相当先进的物流计划能力，在许多案例中表现为全球化过程中的设计能力。

在对中德企业各自的国际化策略进行对比时，我们必须明白它们之间有着完全不同的背景和出发点。大多数德国企业通常是首先感受到海外市场的客户需求，然后随着时间推移自然而然地向海外进行业务拓展，发展国际化物流网络。这种全球业务的拓展往往在一开始并没有一个明确的全球战略，而是一步一步根据需要进行实施，当遇到问题时，则会有针对性地寻找更优的方法进行改进。由于有着多年的积累，这些公司在国际市场拓展方面比中国竞争对手具有更丰富的经验。另一方面，目前具备实力并准备开始全球扩张的中国企业还为数不多。这些企业进入国外市场时，往往采用格林菲尔德策略，这种做法使企业有机会更深层次地分析市场发展，并运用更完整的全球计划来实现国际化步伐。

从中国方面受访者的回馈中，我们可以对中国目前的物流成熟程度得出一个估算结果，这一结果也许会让那些已经从中国企业中获得一些经验的人感到非常吃惊。不过必须声明的是，调研的样本并不能代表所有的中国企业，因为大多数企业目前还只是在国内市场上进行竞争。调查中所涉及的中国企业是那些已经准备或是已经开始国际化运作的企业，因此它们代表的是中国公司的先头部队。这些企业拥有充足的资金，完善的教育培训体系，高水平的领导团队，并经常聘用有海外经验的经理人。因此，不要低估这些即将成为全球竞争者的中国企业。它们随时有可能使全球市场格局发生重大变化。

考虑到在过去的20年来，物流和供应链管理对于企业成功的重要性不断上升，我们可以肯定这些因素在全球化商业活动中将起到更为重要的作用。在任何一种全球化物流策略中都包含三个重要组成部分：第一，在对全球化物流网络进行设计时，必须意识到市场环境是瞬息万变的，并且企业所面临的需求也是不断变化的，高度的柔性和敏捷性是必不可少的。

第二，由于国际化物流网络连接着越来越多的地区和市场，物流需要应对高度的复杂性。因此需要在全球标准化的业务流程和本地适应性之间找到一个平衡。

第三个方面是与物流服务提供商进行战略合作。这种合作所带来的好处已经得到广泛的认可，并且数据显示在国际化环境下，不论是在中国还是德国，企业业务外包程度和范围都是快速增加的。

我们正经历着有史以来进行国际化商业活动和建立全球物流网络最好的时期。当然，这并不是一个简单的任务，企业管理者仍然在各个方面，尤其是物流活动方面，面临着巨大挑战。由于我们的调查问卷中的问题是如此的富有变化和复杂，因此受访者在回答问题中难免会出现自相矛盾的情况。国际化过程不是一个标准化的过程，其相关的问题也不可能完全阐述清楚。实际上，对于实践中的许多问题，我们往往是利用感性认知和实用经验来解决的。

国际化物流管理领域是如此广袤而迷人，我们实在不可能在调查问卷中一一涉及。还有许多问题没有找到答案，但同时也为未来的研究提供了广阔的空间。

参考文献 － 调查报告

Baumgarten, H.; Darkow, I.-L.; Zadek, H. (ed.) (2004): Supply Chain Steuerung und Services: Logistik-Dienstleister managen globale Netzwerke - Best Practices; Published by Springer Verlag; Berlin et al. 2004.

Baumgarten, H.; Herter, M. (1999): Internationalisierung der Logistik; in: Weber, J.; Baumgarten, H. (ed.): Handbuch Logistik 1999 - Management von Material- und Warenflußprozessen; Published by Schäffer Poeschel Verlag; Stuttgart 1999; pp. 828-41.

Beatson, D. (2004): Becoming Truly Global: 3PLs face new Challenges; Source: http://www.inboundlogistics.com/articles/3plline/3plline0104.shtml; Access on March 11, 2007 at 11:37 p.m.

Bowman, R.J. (2006): Globalization, Customer Demand Drive Logistics Outsourcing to the Next Level; in: Global Logistics and Supply Chain Strategies; Vol. 10; No. 6; June 2006; pp. 40-5.

Busse, M. (2002): Transaktionskosten und Wettbewerbspolitik (Transaction Costs and Competition Policy); in: Wirtschaft und Wettbewerb, Vol. 52, Iss. 2, pp. 112-120.

Butts, J. (2006): The Seven Abilities of Highly Effective 3PLs; Source: http://www.inboundlogistics.com/articles/3plline/3plline0906.shtml; Access on March 11, 2007 at 16:02 p.m.

Capgemini; Langley, C.J.; SAP; DHL (ed.) (2006): 2006 Third-Party Logistics - Results and Findings of the 11th Annual Study; Georgia Institute of Technology; Atlanta 2006.

DHL (ed.) (2005): Logistik-Lotse 2005. Leifaden für Logistik - eCommerce - KEP - Spedition - Land-, Luft- und Seeverkehre - Zoll - Außenwirtschaft; 13th, revised edition; Published by Verkehrs-Verlag J. Fischer; Dusseldorf 2005.

Foster, T.A. (2005): Logistics Inside China: The Next Big Supply Chain Challenge; in: Global Logistics and Supply Chain Strategies; Vol. 9; No. 9; September 2005; pp. 28-36.

Foster, T.A.; Armstrong, R. (2006): The Top 25 Global 3PLs: Is Bigger Really Better?; Source: http://www.glscs.com/archives/05.06.25_ 3pls. htm?adcode=90; Access on December 14, 2006 at 17:00 p.m.

Germain, C.; Andrews, S.; Vera, H.; Dunsmore, S. (2006): Worth Repeating – Four Experts outline the Challenges of Global Trade Management; in: Canadian Transportation & Logistics; March 2006; p. 42.

Gershenhorn, A. (2004): The Making of a Successful Global Supply Chain; in: World Trade; Vol. 17; No. 12; December 2004; pp. 19-22.

Gourdin, K.N. (2006): Global Logistics Management: a Competitive Advantage for the 21st Century; 2nd edition; Published by Blackwell Publishing; Oxford 2006.

Harps, L.H. (2003): Global Logistics: Bridging the Culture Divide; Source: http://www.inboundlogistics.com/articles/features/0303_feature01.shtml; Access on December 10, 2006 at 13:30 p.m.

Hoffman, K.C. (2002): Global Growth of 3PLs Is in Response to Customer Demand; Source: http://www.glscs.com/archives/03.02.outsource. htm?adcode=90; Access on December 12, 2006 at 20:15 p.m.

Hofstede, G. (2005): Cultures and Organizations: Software of the Mind; revised and expanded 2nd edition; Published by McGraw-Hill; New York 2005.

Kerr, J. (2006): What's the Right Role for Global 3PLs?; in: Logistics Management; February 2006; pp. 51-4.

Lee, H. L. (2004): The triple-A supply chain; in: Harvard Business Review, Vol. 82 No. 10; pp. 102-112.

Loderhose, B. (2007): Frische Fernöstlich; in: Lebensmittel Zeitung; Vol. 20; May 18, 2007; pp. 32-33.

Ma, S.; Lin, Y; Chen, Z. (2000): Supply Chain Management. China Machine Press, 2000.

MacDonald, A. (2006): Managing a Global vs. Domestic Supply Chain; in: World Trade; Vol. 19; No. 7; July 2006; http://ctl.mit.edu/public/070606_WorldTrade_Sheffi_GlobalvDom.pdf; Access on November 30, 2006 at 21:34 p.m.

Mentzer, J.T.; Myers, M.B.; Stank, T.P. (2006): Handbook of Global Supply Chain Management; Published by Sage Publications; London 2006.

Mercer Management Consulting (ed.) (2005): Mythen der Kontraktlogistik: Mercer Studie untersucht Logistik-Geschäftsmodelle; Source: http://www.einkauf-und-management.at/index.php/einkauf/more/mythen_der_kontraktlogistik; Access on November 10, 2006 at 16:31 p.m.

Nohlen, D. (ed.) (1998): Lexikon Dritte Welt - Länder, Organisationen, Theorien, Begriffe, Personen; completely revised new edition; Published by Rowohlt Taschenbuch Verlag; Reinbek 1998.

Pfohl, H.-C. (2004): Logistiksysteme - Betriebswirtschaftliche Grundlagen; 7th, corrected and updated edition; Published by Springer Verlag; Berlin et al. 2004.

Piszczalski, M. (2002): Global Logistics Issues; in AD&P; January 2002; pp. 16-8.

Richardson, H.L. (2005): How to maximize the Potential of 3PLs; in: Logistics Today; May 2005; pp. 19-21.

Simchi-Levi, D.; Kaminsky, P.; Simchi-Levi, E. (2003): Designing and Managing the Supply Chain: Concepts, Strategies, and Case Studies; Published by McGraw-Hill; 2nd edition; New York 2003.

Straube, F. (2004): e-Logistik - Ganzheitliches Logistikmanagement; Published by Springer Verlag; Berlin et al. 2004.

Straube, F.; Dangelmaier, W.; Günthner, W.A.; Pfohl, H.-C. (2005): Trends und Strategien in der Logistik - Ein Blick auf die Agenda des Logistik-Managements 2010; in: Bundesvereinigung Logistik e.V. (ed.); Published by Deutscher Verkehrs-Verlag; Hamburg 2005.

Transport Intelligence (ed.) (2006): Global Contract Logistics 2006: Outsourcing and Collaboration; Brinkworth, United Kingdom; May 2006.

Tufinkgi, P. (2006): Logistik im Kontext internationaler Katastrophenhilfe; in: Schriftenreihe Logistik der Kühne-Stiftung 9; Published by Haupt Verlag; Bern 2006.

Vijayaraghavan, T. A. S. (2001): Impact of Transportation Infrastructure on Logistics in India; Source: http://www.findarticles.com/p/articles/mi_qa3766/is_200101/ai_n8952924; Access on January 4, 2007 at 18:44 p.m.

Waters, D. (2007): Global Logistics: New Directions in Supply Chain Management, Published by Kogan Page; London 2007.

参考文献 － 插页

AAPA World Port Rankings 2005; Source: http://aapa.files.cms-plus.com/Statistics/WORLD%20PORT%20RANKINGS%202005.xls; Access on August 30, 2007 at 03:40 p.m.

Auswärtiges Amt (2007): Indien: Wirtschaft; Source: http://www.auswaertiges-amt.de/diplo/de/Laenderinformationen/Indien/Wirtschaft.html; Access on June 7, 2007 at 02:45 p.m.

Auswärtiges Amt (2007): Länderinformation: Russische Föderation; Source: http://www.auswaertiges-amt.de/diplo/de/Laender/RussischeFoederation.html; Access on May 2, 2007 at 1:00 p.m.

Auswärtiges Amt (2007): Wirtschaftsdatenblatt Indien; Source: http://www.auswaertiges-amt.de/diplo/de/Laenderinformationen/Indien/Wirtschaftsdatenblatt.html; access on June 5, 2007 11:10 a.m.

Biedermann, D.; Dibenedetto, B. (2007): Twist of Fate; in: Journal of Commerce; New York; Apr. 9, 2007; p. 1.

Bundesagentur für Aussenwirtschaft (2006): Russlands Logistiksektor kann Nachfrage kaum decken; Source: http://www.bfai.de; Access on August 30, 2007 03:20 p.m.

Bundesagentur für Aussenwirtschaft (2006): Transport und Logistk - Russland; Source: http://www.bfai.de; Access on May 03, 2007 02:20 p.m.

Charles, A. (2007): The China Business Handbook 2007; 10th, completely revised and updated edition; Published by Alain Charles Publishing; London 2007.

Chiarello, A. A. (2005): Maersk Logistics USA Inc.; in: Journal of Commerce; New York; Jan 12, 2004; p. 1.

Cygnus Business Consulting and Research (2006): Background note - Logistics; Source: http://www.cygnusindia.com/articles/White%20paper%20on%20CII%20conf%20on%20Logistics%20-Kolkata%2080906.pdf; Access on 30 August, 2007 at 05:05 p.m.

Davies, Tim (2002): Argentina: Crisis? What Crisis?; in: Credit Management; Oct 2002; pp. 35-36.

de Schmidt, A. (2007): Provider am Ganges; in: Logistik Heute 05/2007; pp. 54-55.

de Schmidt, A. (2007): Take off am Taj Mahal; in: Logistik Heute 05/2007; pp. 48-49.

de Schmidt, A. (2007): Zäher Prozess; in: Logistik Heute 05/2007; pp. 56-57.

Dibenedetto, Bill (2007): Growing Pains; Journal of Commerce; New York: May7, 2007; p. 1.

Energy Information Administration (2007): Country Analysis Briefs: Russia; Source: http://www.eia.doe.gov/emeu/cabs/Russia/Background.html; Access on May 03, 2007 at 10:00 a.m.

Federation of Indian Chambers of Commerce and Industry (2007): Current State of Indian Economy; Source: http://www.indiainbusiness.nic.in/indian-economy.pdf; Access on August 30, 2007 at 05:00 p.m.

Franz, M.; Hassler, M. (2007): Eine Milliarde Kunden; in: Logistik Heute 05/2007; pp. 58-60.

Gakenheimer, R. (1997): The urban transport crisis in Europe and North America; in: Journal of American Planning association; Vol. 63; Iss. 4; pp. 530-531.

Gallagher, J. (2007): Racing for NAFTA freights; in: Traffic World, Newark; Mar 12, 2007; p. 1.

Göpfert, I. (ed.) (2006): Logistik der Zukunft – Logistics for the Future; 4th, updated and revised edition; Published by Gabler Verlag; Wiesbaden 2006.

Hausman, W.H.; Lee, H.L.; Subramanian, U. (2005): Global Logistics Indicators, Supply Chain Metrics, and Bilateral Trade Patterns; Source: http://rru.worldbank.org/Documents/Discussions/global_logistics_indicators.pdf; Access on November 22, 2006 at 18:12 p.m.

Hemerling, J.; Michael, D.C.; Michaelis, H. (2006): China's Global Challengers: The Strategic Implications of Chinese Outbound M&A; Published by Boston Consulting Group; May 2006; Source: http://www.bcg.com/publications/files/Chinas_Global_Challengers_May06.pdf; Access on 03 September at 04:45 p.m.

Invest in Germany (2005): Germany: Europe's Logistics Hub – Strengths and Opportunities; Source: http://www.kompetenzcluster.org/fileadmin/vdidaten/Logistik/Fakten/Germany-Logistics-Hub.pdf; Access on August 30, 2007 at 03:35 p.m.

Jahns, C.; Darkow, I.-L.; Weigl, T. (2006): Dynamic Supply Chains in Russia: Industries, Strategies, and Logistics Structures; Published by Bundesvereinigung Logistik e.V. (German Logistics Association); Deutscher Verkehrs-Verlag, Hamburg 2006.

JSC Russian Railways Company Profile (2007); Source: http://www.eng.rzd.ru/page.html?nav_id=145; Access on August 30, 2007 03:30 p.m.

Klaus, P.; Kille, C. (2006): Die TOP 100 der Logistik; 4th edition; Published by Bundesvereinigung Logistik e.V. (German Logistics Association), Deutsche Verkehrs-Zeitung; Deutscher Verkehrs-Verlag, Hamburg 2006.

Lopez Carlos, A. et al. (2006): The Global Competitiveness Index: Identifying the Key Elements of Sustainable Growth; Source: http://www.weforum.org/fweblive/groups/public/documents/wef_member_pdf/gcr_0607_1_1_gcindexes.pdf; Access on August 30, 2007 at 1:00 p.m.

Morton, R. (2006) NAFTA: Twelve years after; in: Logistics Today; Vol. 47, Iss. 2; p. 10.

Open Joint Stock Company Russian Railways (2006): Annual Report 2005; Moscow (2006), Source: http://www.eng.rzd.ru/images/download_im.html?id=3810; Access on August 30, 2007 at 03:45 p.m.

Schulz, J.D. (2006): DHL crashes the party; in: Logistics Management; Vol. 44; Iss. 8; pp. 59-62.

Sissel, K. (1995): Foreign investment picks up; in: Chemical Week. New York; Nov 15, 1995; pp. 30-31.

Smith, James L. (2005): Calculating investment potential in South America; in: World Oil; June 2005; Vol. 216; Iss. 6; pp. 117-221.

Staudenmayer, M. (2007): Hohe Qualität in kleiner Stückzahl; in: Logistik Heute 05/2007; pp. 50-51.

Taylor, J.C., Robideaux, D.R., Jackson, G.C. (2004): U.S.-Canada Transportation and Logistics: Border Impacts and Costs, Causes, and Possible Solutions; in: Transportation Journal; Lock Haven; Fall 2004; Vol. 43, Iss. 4; pp. 5-16.

Verstaen, J.; de Schmidt, A. (2007): Achillesferse mit Schlaglöchern; in: Logistik Heute 05/2007; pp. 52-53.

Waters, D. (2007): Supply Chain Risk Management: Vulnerability and Resilience in Logistics, Published by Kogan Page; London 2007.

World Bank (2005): From Transition To Development: A Country Economic Memorandum for the Russian Federation; Publihed by ABSOLUT Print, Moscow 2005; Source: http://ns.worldbank.org.ru/files/cem/CEM_final_eng.pdf; Access on August 30, 2007 03:00 p.m.

World Bank (2006): Decentralization in India; Source: http://go.worldbank.org/YQWVRQ7N00; Access on May 05, 2007 03:20 p.m.

World Bank (2006): Governance in India; Source: http://go.worldbank.org/
JZ4EGDJQ30; Access on May 03, 2007 03:20 p.m.

World Bank (2007): Growth in India; Source: http://go.worldbank.
org/1DLEBZ7C50; Access on May 04, 2007 02:40 p.m.

World Bank (2007): India at a glance; Source: http://devdata.worldbank.
org/AAG/ind_aag.pdf; Access on 30 August, 2007 at 05:00 p.m.

World Bank (2007): India Foreign Trade Policy; Source: http://
go.worldbank.org/RJEB2JGTC0; Access on August 30, 2007 at 04:30 p.m.

World Bank (2007): India Transport Sector; Source: http://go.worldbank.
org/FUE8JM6E40; Access on August 30, 2007 at 04:30 p.m.

World Trade Organization (2007): Accessions: Russian Federation; Source:
http://www.wto.org/english/thewto_e/acc_e/a1_russie_e.htm; Access on August 30, 2007 03:00 p.m.

WTO (2006): WTO International Trade Statistics, Source: http://www.wto.
org/english/res_e/statis_e/its2006_e/its06_toc_e.htm; Access on August
30, 2007 at 03:55 p.m.